AF389166

Better Animal Feeding for Improving the Quality of Ruminant Meat and Dairy

Better Animal Feeding for Improving the Quality of Ruminant Meat and Dairy

Editors

Manuel Delgado-Pertíñez
Alberto Horcada

MDPI • Basel • Beijing • Wuhan • Barcelona • Belgrade • Manchester • Tokyo • Cluj • Tianjin

Editors
Manuel Delgado-Pertíñez
Universidad de Sevilla. Ctra.
Utrera
Spain

Alberto Horcada
Universidad de Sevilla. Ctra.
Utrera
Spain

Editorial Office
MDPI
St. Alban-Anlage 66
4052 Basel, Switzerland

This is a reprint of articles from the Special Issue published online in the open access journal *Foods* (ISSN 2304-8158) (available at: https://www.mdpi.com/journal/foods/special_issues/Ruminant_Meat_Dairy).

For citation purposes, cite each article independently as indicated on the article page online and as indicated below:

LastName, A.A.; LastName, B.B.; LastName, C.C. Article Title. *Journal Name* **Year**, *Volume Number*, Page Range.

ISBN 978-3-0365-4317-8 (Hbk)
ISBN 978-3-0365-4318-5 (PDF)

Contents

About the Editors

Manuel Delgado Pertíñez

Senior Lecturer at the School of Agricultural Engineers, University of Seville, graduated with a PhD in Veterinary Science from the University of Córdoba in 1994. His main research lines focus on animal nutrition-quality of products (meat and dairy), especially in Mediterranean pastoral- based farming systems and the use of agro-industrial by-products, to promote the sustainability and resilience of ecosystems and the circular economy. He is the recipient of a range of fellowships (FPI-Training of Research of the Ministry of Science and Innovation-Government of Spain; Community program for Research and Technological Development of the EU; Grants for the Improvement of Research Staff of the Government of Andalusia) and has made several research and development stays (Hassan II Institute of Agronomy & Veterinary Medicine, Rabat-Morocco; Rowett Research Institute, Aberdeen-United Kingdom; National Autonomous University of Mexico). He is currently the Academic Coordinator of the PIMA Network "Sustainable Agriculture" of the Organization of Ibero-American states (OEI).

Alberto Horcada

Senior Lecturer at the Hight School of Agricultural Engineers, University of Seville (Spain), graduated with a PhD from the Public University of Navarre in 1996 after completing formation in Biology and Veterinary Sciences. His current research interests is Animal Production Sciences including meat quality products and development of local livestock breeds in mediterranenan areas of Europe. He is member of New technologies for animal improvement and their production systems group (MERAGEM, PAIDI AGR-273) in Andalucía (Spain). He is the recipient of a range of fellowships, including the FPI pre-Doctoral Fellowship (1992–1995) to university teacher training. Before his activity as Senior Lecturer, he has carry out his professional activity as a consultant for the creation of the PGI quality label for beef in Spain.

Editorial

Better Animal Feeding for Improving the Quality of Ruminant Meat and Dairy

Manuel Delgado-Pertíñez * and Alberto Horcada

Departamento de Ciencias Agroforestales, Escuela Técnica Superior de Ingeniería Agronómica, Universidad de Sevilla, Ctra. Utrera km 1, 41013 Sevilla, Spain; albertohi@us.es
* Correspondence: pertinez@us.es; Tel.: +34-954-486-449

Citation: Delgado-Pertíñez, M.; Horcada, A. Better Animal Feeding for Improving the Quality of Ruminant Meat and Dairy. *Foods* **2021**, *10*, 1076. https://doi.org/10.3390/foods10051076

Received: 8 May 2021
Accepted: 11 May 2021
Published: 13 May 2021

Publisher's Note: MDPI stays neutral with regard to jurisdictional claims in published maps and institutional affiliations.

The quality of meat and dairy products can be evaluated from the perspective of the farmer seeking high yields and profits or the consumer for whom sensory characteristics are the most important, although health and ethical aspects, such as animal welfare and the environmental impact of the production system, are increasingly becoming concerns worldwide. Animal nutrition is one of the most important environmental factors that significantly influences the quality of meat, milk, and other dairy products; therefore, emphasis is often placed on improving the quality of feed. A main target for improving the nutritional characteristics of meat and dairy is the enhancement of lipid quality, which can be achieved by increasing the content and improving the composition of beneficial fatty acids (FAs)—n-3 polyunsaturated FA (PUFA) and conjugated linoleic acid (CLA)—and decreasing the n-6:n-3 PUFA ratio. Factors such as the forage:concentrate ratio, dietary fat supplements, pasture, etc. have a crucial effect on dairy and meat quality. Some studies have shown that meat and dairy from ruminants grazing on pasture are enriched with bioactive substances of natural origin like phenolic compounds, fat soluble vitamins, terpenes, and lipid components. These animals are also able to consume increasing amounts of by-products or 'unconventional' animal feedstuffs, which can improve the health-giving properties of products. In addition, dietary manipulations favouring polyunsaturated FA in dairy and meat lipids increase the risk of lipoperoxidation, which can be efficiently prevented by the use of antioxidants in the diet. Furthermore, the search for biomarkers that link the composition of animal products to livestock feed has become a target of scientific research; these biomarkers allow for traceability from farm to fork, based on the herbivore diet and geographical origin. In this context, the Special Issue 'Better Animal Feeding for Improving the Quality of Ruminant Meat and Dairy' aims to provide an integrated analysis of the major aspects of the nature and composition of ruminant meat and dairy (i.e., FA profile, antioxidants, vitamins, muscle:fat ratio, flavor, etc.) and the effect of better animal feeding on the improvement of nutritional and sensory qualities and functional properties beneficial to human health. This Special Issue comprises seven valuable works of original research on product quality; four studies were conducted on pasture-based systems (two on sheep cheese quality, one on cow's milk quality, and one on beef quality), two on alternative feedstuff-based systems (goat dairy product quality), and one on conventional concentrate-fed animals (beef quality).

Of the articles on pasture-based systems, three compared grass-fed regimes with conventional concentrate-fed regimes. Serrapica et al.'s [1] study on a traditional pecorino cheese associated with management and feeding-system seasonality was conducted in two mountain dairy sheep farms rearing the native Bagnolese breed using an outdoor, pasture-based system from April to October (fed with pasture integrated with hay and concentrate mixture) and an indoor system during the rest of the year (fed with hay and cereal grains). The pasture-based (outdoor management system) cheese had higher percentages of unsaturated FAs (UFA), C18:3, and rumenic (CLA cis-9, trans-11) and vaccenic (C18:1 trans-11) acids, and lower percentages of C14:0 and C16:0, along with a reduction in the atherogenic index. The pasture-based cheeses also displayed higher intensities of almost all sensory attributes, including odour, flavour, taste, and texture descriptors. Optimisation of farm production includes the rational use of forages and respect for the environment, and offers the best quality of products to consumers. Therefore, De la Torre-Santos et al. [2] investigated the effect of the mode of supply of grass to dairy cows (grazing, cut-and-carry, zero-grazing, or grass silage) on milk performance and milk antioxidant and FA profiles to identify biomarkers of the feeding systems. Changes in milk yield and composition were associated with the feeding system. The grazing system promoted a higher dry matter intake and greater milk yield as well as a higher proportion of UFA (with significant differences in the proportion of vaccenic and rumenic acids) and lutein than the grass silage and zero-grazing systems. The authors propose the 18:1 trans-11 to 18:1 trans-10 ratio as a biomarker to identify the milk produced by cattle in the grazing system. The third study [3] comparing feeding regimes investigated beef quality. The primary aim of this study was to determine the influence of two organic production systems—organic grass-fed and organic concentrate-fed—compared to a conventional concentrate-fed system, on the evolution of texture properties and histological attributes of muscle fibres during the ageing process of the meat of the Retinta breed. Although the meat of organic grass-fed animals was initially tougher, its speed of tenderisation was higher in the first ageing days than that of meat from concentrate-fed systems. In all systems of feeding, sarcomere length increased during the ageing period, and showed a negative correlation with shear force, which is related to the tenderising of the meat. Finally, the fourth study on pasture-based systems [4] is the first nutritional analysis of dairy products from the traditional Roja Mallorquina sheep breed raised on a part-time grazing system, which could support the implementation of strategies to promote their commercialisation and obtain product labelling as 'pasture-fed' or other specific marks. Results showed that fresh soft cheese, compared to the original milk and ripened cheese, was generally characterised by better nutritional value for human health according to the fat-soluble components—a favourable level of retention of retinol and α-tocopherol and a lower percentage of saturated FAs (SFA) and lower atherogenic (AI) and thrombogenic (TI) indices. Furthermore, acetoin and products of lactose and citrate metabolism played an important role in the development of the aromatic attributes of both kinds of cheese.

In the Mediterranean region, high amounts of by-products are available that can be used as alternative feedstuffs for ruminants, lowering feed costs, and enhancing farm sustainability, while reducing the environmental impact of livestock production. Monllor et al. [5] determined the effect of three levels of inclusion of by-product silages (artichoke plants and broccoli by-product) in goat diets on milk yield and quality. They proposed that the threshold level of inclusion of by-products in diets, without negative effects on milk yield and composition and the metabolic status of the animals, would be 40% of the dietary dry matter. From the point of view of human health (AI and TI), the study demonstrated that the inclusion of artichoke plant silage in the animals' diet improved the milk lipid profile compared to broccoli silage, due to a lower SFA content (C12:0, C14:0, and C16:0) and a higher PUFA concentration, especially of vaccenic and rumenic acids, without any differences from the control treatment. Spain is the primary producer of oranges in the European Union and the sixth largest global producer; orange pulp, the principal by-product, can partially replace cereal grains in ruminant feedstuffs. Therefore, Guzmán et al. [6] evaluated the influence of diets with dry orange pulp pellets on the physicochemical characteristics, sensory properties, and volatile compound profiles of cheeses traditionally made from the milk of the Payoya breed of goat. Results showed that dried citrus pulp can be used in the goat diet as a substitute for cereal in concentrates because it did not substantially affect the distinctive final characteristics of the ripened raw milk cheeses.

In the last paper of this Special Issue, Barahona et al. [7] studied the effect of two feeding systems based on concentrate and maize silage and two packaging systems, vacuum and modified atmosphere, on the quality of meat from the Avileña breed of cattle. Their results showed that animals fed with concentrate had higher carcass weight; however, the use of maize silage improved the tenderness of meat and the FA content. In general, vacuum-packed meat from maize silage-fed animals was the most preferred by consumers. An alternative feeding system based on the use of maize silage may improve some aspects of meat quality.

In summary, the Special Issue provides evidence that production systems based on grass and forages produce healthier products with better sensory attributes, thereby promoting the consumption of healthier foods. At the same time, as suggested by several authors, high amounts of by-products can be used as alternative feedstuffs for ruminants, enhancing farm sustainability while improving the health-giving properties of products.

Author Contributions: M.D.-P. and A.H. contributed equally to the writing and editing of the specific editorial note. All authors have read and agreed to the published version of the manuscript.

Funding: This research received no external funding.

Conflicts of Interest: The authors declare no conflict of interest.

References

1. Serrapica, F.; Masucci, F.; Di Francia, A.; Napolitano, F.; Braghieri, A.; Esposito, G.; Romano, R. Seasonal Variation of Chemical Composition, Fatty Acid Profile, and Sensory Properties of a Mountain Pecorino Cheese. *Foods* **2020**, *9*, 1091. [CrossRef] [PubMed]
2. De La Torre-Santos, S.; Royo, L.J.; Martínez-Fernández, A.; Chocarro, C.; Vicente, F. The Mode of Grass Supply to Dairy Cows Impacts on Fatty Acid and Antioxidant Profile of Milk. *Foods* **2020**, *9*, 1256. [CrossRef] [PubMed]
3. García-Torres, S.; López-Gajardo, A.; Tejerina, D.; Prior, E.; de Vaca, M.C.; Horcada, A. Effect of Two Organic Production Strategies and Ageing Time on Textural Characteristics of Beef from the Retinta Breed. *Foods* **2020**, *9*, 1417. [CrossRef] [PubMed]
4. Gutiérrez-Peña, R.; Avilés, C.; Galán-Soldevilla, H.; Polvillo, O.; Ruiz Pérez-Cacho, P.; Guzmán, J.L.; Horcada, A.; Delgado-Pertíñez, M. Physicochemical Composition, Antioxidant Status, Fatty Acid Profile, and Volatile Compounds of Milk and Fresh and Ripened Ewes' Cheese from a Sustainable Part-Time Grazing System. *Foods* **2021**, *10*, 80. [CrossRef] [PubMed]
5. Monllor, P.; Romero, G.; Atzori, A.S.; Sandoval-Castro, C.A.; Ayala-Burgos, A.J.; Roca, A.; Sendra, E.; Díaz, J.R. Composition, Mineral and Fatty Acid Profiles of Milk from Goats Fed with Different Proportions of Broccoli and Artichoke Plant By-Products. *Foods* **2020**, *9*, 700. [CrossRef] [PubMed]
6. Guzmán, J.L.; Delgado Pertíñez, M.; Galán Soldevilla, H.; Ruiz Pérez-Cacho, P.; Polvillo Polo, O.; Zarazaga, L.Á.; Avilés Ramírez, C. Effect of Citrus By-product on Physicochemical Parameters, Sensory Analysis and Volatile Composition of Different Kinds of Cheese from Raw Goat Milk. *Foods* **2020**, *9*, 1420. [CrossRef] [PubMed]
7. Barahona, M.; Hachemi, M.A.; Olleta, J.L.; González, M.d.M.; Campo, M.d.M. Feeding, Muscle and Packaging Effects on Meat Quality and Consumer Acceptability of Avileña-Negra Ibérica Beef. *Foods* **2020**, *9*, 853. [CrossRef] [PubMed]

 *foods*

Article

Physicochemical Composition, Antioxidant Status, Fatty Acid Profile, and Volatile Compounds of Milk and Fresh and Ripened Ewes' Cheese from a Sustainable Part-Time Grazing System

Rosario Gutiérrez-Peña [1], Carmen Avilés [2], Hortensia Galán-Soldevilla [2], Oliva Polvillo [3], Pilar Ruiz Pérez-Cacho [2], José Luis Guzmán [4], Alberto Horcada [1] and Manuel Delgado-Pertíñez [1,*]

[1] Departamento de Ciencias Agroforestales, Escuela Técnica Superior de Ingeniería Agronómica, Universidad de Sevilla, 41013 Sevilla, Spain; charo-84@hotmail.com (R.G.-P.); albertohi@us.es (A.H.)

[2] Departamento de Bromatología y Tecnología de los Alimentos, Campus de Rabanales, Universidad de Córdoba, 14071 Córdoba, Spain; v92avrac@uco.es (C.A.); bt1gasoh@uco.es (H.G.-S.); pilar.ruiz@uco.es (P.R.P.-C.)

[3] Servicio General de Investigación Agraria, Escuela Técnica Superior de Ingeniería Agronómica, Universidad de Sevilla, 41013 Sevilla, Spain; oppolo@us.es

[4] Departamento de Ciencias Agroforestales, Escuela Técnica Superior de Ingeniería, 'Campus de Excelencia Internacional Agroalimentario, ceiA3' Campus Universitario de la Rábida, Carretera de Huelva-Palos de la Frontera s/n., Universidad de Huelva, 21819 Huelva, Spain; guzman@uhu.es

* Correspondence: pertinez@us.es; Tel.: +34954486449

Citation: Gutiérrez-Peña, R.; Avilés, C.; Galán-Soldevilla, H.; Polvillo, O.; Ruiz Pérez-Cacho, P.; Guzmán, J.L.; Horcada, A.; Delgado-Pertíñez, M. Physicochemical Composition, Antioxidant Status, Fatty Acid Profile, and Volatile Compounds of Milk and Fresh and Ripened Ewes' Cheese from a Sustainable Part-Time Grazing System. *Foods* **2021**, *10*, 80. https://doi.org/10.3390/foods10010080

Received: 8 December 2020
Accepted: 31 December 2020
Published: 3 January 2021

Publisher's Note: MDPI stays neutral with regard to jurisdictional claims in published maps and institutional affiliations.

Abstract: We conducted the first nutritional analysis of dairy products from the traditional Roja Mallorquina sheep breed. Samples of bulk raw milk were taken twice a month from December 2015 to March 2016 from sheep fed using a part-time grazing system, and fresh soft (FC, $n = 8$) and ripened (RC, $n = 8$) cheeses were made. The variability in vitamins, total phenolic compounds (TPC), total antioxidant capacity (TAC), and fatty acid (FA) content was influenced by the cheese-making process (differences between the cheese and the original milk) and by the type of cheese-making technology (mainly related to heating, the use of starter culture, and ripening). The most notable physicochemical characteristic of the cheeses was low fat content (24.1 and 29.6 g/100 g for FC and RC). Milk and RC were characterised by major concentrations of retinol (211.4 and 233.6 µg/100 g dry matter (DM), respectively) and TPC (18.7 and 54.6 µg/100 g DM, respectively), while FC was characterised by major concentrations of retinol (376.4 µg) and α-tocopherol (361.7 µg). The fat-soluble components of the FC generally exhibited better nutritional value for human health than those of the milk and RC, with a higher level of retinol and α-tocopherol; lower values for saturated FA, atherogenic, and thrombogenic indices; and higher levels of monounsaturated FA, polyunsaturated FA, n-3, and n-6. Acids, alcohols, and ketones comprised almost 95% of the volatile compounds detected. Acetoin and products of lactose and citrate metabolism played an important role in the development of the aromatic attributes of both kinds of cheese. This preliminary study can contribute to add value to these traditional products according to healthy nutritional criteria and supports the implementation of strategies to promote their commercialisation and obtain product labelling as "pasture-fed" or specific marks.

Keywords: antioxidant capacity; dairy product quality; n-3 and n-6 fatty acids; retinol; Roja Mallorquina sheep; tocopherol; total phenolic compounds; volatile compounds

1. Introduction

The disappearance of many pastoral farming methods in Europe has revealed the importance of sustainable livestock management for environmental conservation [1]. In the Balearic Islands (Spain), traditional sheep systems based on the use of autochthonous breeds deserve to be highlighted. Many of these breeds, such as the Roja Mallorquina, are considered endangered according to the Official Catalogue of Spanish Livestock Breeds [2]. The Roja Mallorquina sheep is a meat breed raised under grazing systems. However, traditionally, some breeders have produced artisanal Mallorquin cheeses with a highly characteristic flavour attributed to the breed's very fatty milk. Normally, two types of cheese are made: a fresh soft cheese (white in the beginning and slightly yellowish after ten or fifteen days) and a cheese ripened for approximately two months [3]. To our knowledge, no scientific data have been published so far concerning milk or fresh and ripened cheeses from the Roja Mallorquina.

Cheeses made from small ruminants' milk are widely appreciated for their organoleptic properties. The composition of the lipids plays an essential role in the sensory traits of these products. However, dairy products have often been associated with many negative health effects due to their saturated fatty acid (SFA) content; this may lead to increased low-density lipoprotein cholesterol (LDL) levels and thus an increased risk of cardiovascular disease (CVD) [4]. Nevertheless, recent research has demonstrated the benefits of full-fat dairy consumption [5] and suggests that milk has a neutral effect on cardiovascular health, whereas fermented milk and cheese may be beneficial. Furthermore, foods containing natural antioxidants have grown in popularity, as antioxidants can neutralise harmful effects of free radicals both in humans and animals and because oxidation processes in milk can result in a deterioration of its nutritional quality [6,7]. However, few studies have investigated the milk and cheese characteristics of endangered sheep breeds.

Moreover, there is interest in assessing the effects of different feeding regimes on dairy quality, including the pathways that transfer nutritional components from milk to cheese and other dairy products. On the one hand, several feeding strategies, including grass feeding and pasture grazing, are known to confer specific organoleptic features on the dairy products while improving nutrition [8]. Moreover, in some European countries (including Spain) and mainly in cow dairy (most likely due to their large volume and economic importance), industry or marketing intermediaries fund farmers more for milk produced in grassland conditions [9–11]. Although the small ruminant dairy sector is more limited [10] and, in Spain, incentive payments for pasture-fed milk do not yet exist, labelling products as "pasture-fed" could improve the profitability of farms and industries. On the other hand, while numerous studies have been carried out on FA transfer, results vary: some authors have suggested that the FA profile of fresh or ripening cheese reflects that of the milk from which it has been made [12–14], while others [15–17] have suggested that alterations to the FA profile occur during ripening. However, few studies have specifically addressed the transfer of fat-soluble vitamins from milk to cheese [18], and no reports are available on phenolic compounds. Therefore, the aims of the present work were (i) to perform a nutritional analysis of the dairy products (milk, fresh and ripened cheeses) of the Roja Mallorquina sheep breed, including the determination of their physicochemical, antioxidant, fatty acid (FA), and volatile properties and (ii) to identify and explain changes in nutritional composition during the manufacturing procedure.

2. Materials and Methods

2.1. Study Area, Experimental Farm, and Feeding Management

The study was conducted on Mallorca (Balearic Islands) in the Mediterranean Sea in eastern Spain. A farm of indigenous Roja Mallorquina ewes was used in the present study. There are 55 farms included in the Breed registry of the Roja Mallorquina, with an average size of 73 adult animals per farm [2]. It is a rustic breed, and it is a good dam with good dairy conditions. The reproductive management of farms is discontinuous. Males are introduced into the female herd between May and November. Therefore, the lambing period is concentrated between September and April, coinciding with maximum forage production.

The management system is fundamentally based on grazing. Cultivated pastures, such as oats (*Avena sativa*), ryegrass (*Lolium* sp.), and barley (*Hordeum vulgare*), are used for animal feed in the form of green forage (direct grazing or cut grass), dry or ensiled forage, grains, and stubble. The natural grazing area is characterised by pastures with low arboreal stratum density of pine trees (mainly, *Pinus halepensis*) that allow the development of the herbaceous layer (most commonly *Dactylis glomerata*, *Brachypodium retusum*, and *Ampelodesmos mauritanica*) and the shrub layer (*Pistacia lentiscus*, *Rhamnus alaternus*, *Erica multiflora*, and various *Cistus* species) [19]. Natural grass surfaces are found to a lesser extent (most commonly *Dactylis glomerata*, *Hyparrhenia hirta*, and *Brachypodium retusum*) [19]. In winter and spring, the sheep graze on natural and cultivated pastures, while in summer and early autumn, they take advantage of stubble and graze on wooded areas. To meet the nutritional requirements of late pregnancy and lactation, ewes consume purchased grains (0.9 kg/ewe and day) and self-produced forage (from 0.3 to 1.1 kg/ewe and day, depending on pasture yield). Samples from all the feeds supplied to the ewes were collected, and their chemical composition was analysed (Table 1).

Table 1. Chemical composition of the main supplementary feeds during lactation.

Item	Chemical Composition (% Dry Matter, DM)				
	DM	Crude Protein	Crude Fibre	Ether Extract	Ash
Husks and broken cereals	88.8	17.4	8.50	2.24	6.57
Cereal mix (barley, 40%; oats, 40%; beans, 20%)	89.8	13.3	10.3	2.94	2.58
Oat hay	80.6	6.04	40.2	2.40	6.62
Barley straw	91.7	3.70	38.2	1.62	7.63
Oat silage	30.3	8.32	29.6	3.36	7.57

2.2. Milk Collection, Cheese Manufacture and Sampling

After weaning the lambs in November, the dams were milked with an automated parallel milking machine once a day for cheese manufacture from December 2015 to March 2016 (approximately from the middle to the end of lactation). The dams were included in the milking programme according to when their lambs were weaned, and they were removed once they were dried off, but the number of milking ewes remained stable over these months (an average of 150 animals). After homogenising the total yield of daily milk using a mechanical shaker, bulk raw milk samples were taken twice a month ($n = 8$, four replicate samples for each stage of lactation) for cheese-making. Five aliquots of whole milk from each sample were placed in 50 mL plastic bottles for chemical analyses. The sheepherder made cheese in two vats with the same bulk milk sample, one to make fresh soft cheese (FC, $n = 8$) and the other to made ripened cheese (RC, $n = 8$). The cheeses were made following traditional manufacturing conditions. Overall, 16 cheese-making processes were completed. The FC was made with pasteurised milk and without adding starter culture. Milk was pasteurised at 63 °C for 30 min and cooled to the ripening temperature (34 °C). Calf rennet commonly used by farmers (>60% chymosin and <40% pepsin) was added in the amount of 1 mL per 3 L of milk, following the manufacturer's instructions, to obtain clotting within 45 min. After coagulation, all the curds were cut to obtain grains the size of millet. Salt (NaCl) was added to this mixture in a proportion of 1%, w/v. The curd was stirred manually for approximately 20 min, and then the whey was drained. The curd was packed into cylindrical moulds (500 g), and the cheeses were transferred to an airing chamber at a temperature of 3–4 °C without humidity control. The cheeses were sold the next day with a shelf life of 8 days in refrigeration.

The RC was made with starter culture added to unpasteurised milk. After heating to 34 °C, the ferment was added in the proportion of 6 g per 100 L of milk (Mesophilic Starter Series MA 4001, Dupont Nutrition and Biosciences Ibérica, Valencia, Spain). After 30 min of ripening, calf rennet was added in the same proportion to obtain clotting in 45 min. After coagulation, the curd was cut with a lyre of parallel wires to obtain grains the size of hazelnuts. The curd was stirred semi-mechanically for 30–40 min, and then the whey was drained. The curd was packed into cylindrical moulds (500 g or 1 kg) and pressed in a hydraulic press (2.4 kPa) for 1 h. Then, they were salted by brining in NaCl solution (6 °C, 16° Bau, and pH 5.15–5.20) for 8 h. Afterwards, the cheeses were transferred to an airing chamber for 72 h at 12 °C and 80% relative humidity. Finally, the cheeses were transferred to a maturation chamber for 60–70 days of storage at 13 °C and 85–90% relative humidity.

The milk samples and cheeses were transported in a portable cooler with ice. Once in the laboratory, the cheeses were cut into sections (approximately 200 g each) and vacuum-packed. All milk and cheese samples were wrapped in aluminium sheets to preserve them from light and frozen at −20 °C until analysis, except the samples for the analysis of vitamins, which were frozen at −80 °C.

2.3. Chemical Analyses

Before analysis, the feed samples were dried and ground using a Wiley mill with a 1 mm screen. AOAC methods [20] were used to determine dry matter (DM, method 934.01), crude protein (method 984.13), crude fibre (method 978.10), and ash (method 942.05) content.

The milk's basic chemical composition (DM, fat, protein, and lactose) was analysed with a Milkoskan FT (Fourier transform infrared analysis) in a CombiFoss 5000 (Foss Electric, Hillerød, Denmark) calibrated against known standards and subjected to quality controls and inter-comparative trials [21]. In cheese, the fat [22], DM [23], ash, and protein [24] contents were measured. For the determination of nitrogen content, the Kjeldahl method was used, and the total nitrogen content was multiplied by 6.38 to determine total protein. The sodium chloride content was analysed using a back titration with potassium thiocyanate to determine the concentration of chloride ions in the solution (the Volhard method) [23]. The pH was measured with a pH meter (pHmetro HANNA FHT-803) with a pH electrode, according to de la Haba et al. [25].

The procedures described by Guzmán et al. [26] were used to measure FA composition. Fat extraction from 0.1 g of freeze-dried milk or cheese and the direct methylation of FAs were performed in a single-step procedure based on the method by Sukhija and Palmquist [27] and revised by Juárez et al. [28]. Separation and quantification of FA methyl esters (FAMEs) were carried out using a gas chromatograph (Agilent 6890N Network GS System, Agilent, Santa Clara, CA, USA) equipped with a flame ionisation detector (FID) and automatic sample injector HP 7683, and fitted with an HP-88 J&W fused silica capillary column (100 m, 0.25 mm i.d., 0.2 μm film thickness; Agilent Technologies Spain, S.L., Madrid, Spain). Nonanoic acid methyl ester (C9:0 ME) was used as an internal standard (Sigma Aldrich Co., Madrid, Spain). Individual FAs were identified by comparing their retention times with those of the authenticated standard FA mix Supelco 37 (Sigma). The CLA (conjugated linoleic acid) isomers (*cis*-9, *trans*-11 and *trans*-10, *cis*-12) were identified by comparing retention times with those of another authenticated standard (Matreya, LLC, Pleasant Gap, USA).

Fat-soluble vitamin (A and E) analyses of 1.5–2 mL of milk or 2 g of cheese from each relevant sample were based on the methods by Herrero-Barbudo et al. [29] and Chauveau-Duriot et al. [30] and modified by Gutiérrez-Peña et al. [31]. Chromatographic analysis was carried out on an Acquity UPLC, with a fluorimetric detector and isocratic pump, PDA and 150 × 2.1 mm Acquity UPLC HSS T3 1.8 μm column (Waters, Saint-Quentin-en-Yvelines, France). Tocopherols and retinol were positively identified by comparing their retention times with those of high-purity standards of the measured substances (all-*trans*-retinol, α-tocopherol, β-tocopherol, and γ-tocopherol; Sigma Aldrich Co., Madrid, Spain). Other standards of high purity (retinyl acetate, retinyl palmitate, and tocopheryl acetate; Sigma Aldrich Co., Madrid, Spain) were used as internal standards. Since vitamins A and E are fat-soluble, it was necessary to normalise their concentrations against fat in order to determine how milk composition and the cheese-making process contributed to their variation among cheese samples [18].

Total antioxidant capacity (TAC) in milk was analysed by the ABTS method (2,2′-azino-bis (3-ethylbenzothiazoline-6-sulphonic acid)) of Pellegrini et al. [32], modified by Delgado-Pertíñez et al. [33]. Before any measurements were taken for milk, samples were sonicated for 10 min and diluted 10–20 times. Cheese samples for analysis were prepared according to the procedure of Revilla et al. [34] by diluting 2.5 mg of ground cheese in 10 mL of water. After stirring in a water bath at 40 °C, the mixture was centrifuged at 20 °C for 30 min ($3000 \times g$). The supernatant was recovered and brought up to a final volume of 10 mL, and a suitable amount of sample (20 µL) was used to measure TAC as in milk. The water-soluble vitamin E analogue Trolox was used as a standard, and TAC was expressed as mmol Trolox equivalents.

Total phenolic compounds (TPC) in milk were quantified according to the procedure described by Vázquez et al. [35] using the Folin–Ciocalteu method as modified by Guzmán et al. [26]. Cheese samples for analysis were prepared by diluting 10 g of ground cheese in 10 mL of methanol, and after vortexing for 40 min, the mixture was centrifuged at 5 °C for 10 min ($3000 \times g$). The supernatant was recovered, and then the same procedure as for milk was followed. Standard solutions of gallic acid (GA) were used to express the phenolic compounds as g of GA equivalents.

Following Guzmán et al. [36], 5 g of cheese samples were processed in order to extract their volatile compounds with solid-phase microextraction (SPME). A 1 cm long × 110 µm diameter divinylbenzene/carboxen/polydimethylsiloxane fibre (DVB-CAR-PDMS; Supelco, Bellefonte, PA, USA) was fixed in the headspace of the vial. Volatile compound analysis by gas chromatography-mass spectrometry (GC-MS) was also performed following Guzmán et al. [36]. The volatile compounds were tentatively identified by comparing their retention index (obtained using a series of n-alkanes in diethyl ether analysed under the same conditions) to those previously described in the literature and/or comparing their mass spectra with those in the National Institute of Standards and Technology library (NIST; Gaithersburg, MD, USA).

All chemical determinations were made in duplicate.

2.4. Data Processing and Statistical Analysis

The data on the nutritional composition of the three dairy products collected from each of the eight cheese-making sessions (four in each phase of lactation) were analysed using IBM SPSS Statistics for Windows (version 25.0; IBM Corp., Armonk, NY, USA). The normality of variables was assessed using descriptive statistics for asymmetry and kurtosis and the homogeneity of the variances was estimated with Levene's test. To meet the objectives of this study, we used contrasts to investigate the following phenomena. (i) The first was the effects of manufacturing procedures between original milk and cheese (M vs. FC; M vs. RC) on fat-soluble vitamins, TPC, TAC, and FA composition (contents, proportions, categories, and indices). The data were analysed with the repeated measures procedure, and the model included fixed within-subjects factors for the product and lactation stage (repeated measures) as well as the interactions between these factors; the cheese-making session was the replicate. (ii) The second was differences among the fresh and ripened cheeses (FC vs. RC) on fat-soluble vitamins, TPC, TAC, and FA composition. The data were analysed with the repeated measures procedure, and the model included a fixed between-subjects factor for product and a fixed within-subject factor for lactation stage (repeated measures), as well as the interactions between these factors; cheese-making session was the replicate.

Only the product effect means are presented for both types of contrast, as the lactation stage effect and the interaction between product and lactation stage effects were not significant for most of the parameters analysed ($p > 0.05$). Furthermore, Pearson correlation coefficients were calculated for some of the variables used. Statistically significant differences and trends were defined as $p \leq 0.05$ and $0.05 < p \leq 0.10$, respectively.

For the volatile compound content, a multivariate analysis was performed with the principal component analysis (PCA) command of the XLSTAT software (Addinsoft Inc., New York, NY, USA). The results were transformed by an orthogonal rotation (Varimax) with Kaiser normalisation (three components extracted).

3. Results

3.1. Physicochemical Composition and Antioxidant Status

Table 2 presents the means for the physicochemical parameters of fresh and ripened Roja Mallorquina ewes' cheese and the corresponding original milk. The DM, fat, and protein contents of the milk ranges were 18.3–20.7, 6.58–8.72, and 6.08–7.31 g/100 g milk, respectively; the contents of FC ranges were 46.1–48.9, 22.2–25.0, and 11.9–17.1 g/100 g cheese, respectively; and the RC contents ranges were 62.2–65.5, 27.0–33.0, and 22.8–28.0 g/100 g cheese, respectively. The fat/DM ranges were 48.2–52.5 and from 41.2–52.5 g/100 g DM for FC and RC, respectively. Finally, the ash (g/100 g cheese), NaCl (g/100 g cheese), and pH ranges were 1.57–1.92, 0.88–1.54, and 6.43–6.65 for FC and from 3.04–4.14, 0.84–1.38, and 5.17–5.39 for RC.

Table 2. Physicochemical composition of fresh and ripened Roja Mallorquina ewes' cheese and corresponding original milk.

Item	Milk ($n = 8$)		Fresh Cheese ($n = 8$)		Ripened Cheese ($n = 8$)	
	Mean	SEM	Mean	SEM	Mean	SEM
Dry matter (DM) (%)	19.5	0.39	47.5	1.3	64.2	0.78
Lactose (%)	4.54	0.06				
Fat (%)	7.47	0.31	24.1	1.2	29.6	2.4
Protein (%)	6.57	0.18	14.6	0.34	25.6	0.33
Fat/DM (%)			50.7	1.5	46.2	4.7
Ash (%)			1.76	0.03	3.58	0.08
NaCl (%)			1.16	0.09	1.04	0.07
pH			6.54	0.03	5.27	0.03

Table 3 shows the fat-soluble vitamins, TPC, and TAC of the dairy products. The retinol and α-tocopherol contents of milk ranges were 113.4–281.2 and 53.1–156.8 µg/100 g DM, respectively; the contents of FC ranges were 218.9–520.7 and 208.5–549.3 µg/100 g DM, respectively; and the RC contents ranges were 180.0–310.1 and 11.7–49.9 µg/100 g DM, respectively. The TPC content of milk, FC, and RC ranges were 16.8–20.9, 4.0–8.0, and 34.8–69.8 mg GA (gallic acid) equivalents/100 g DM, respectively. The TAC content for milk, FC, and RC ranges were 7.4–50.4, 61.8–139.1, and 56.7–138.9 µmol Trolox equivalents/g DM, respectively. All these components were affected by cheese manufacturing. The fat-soluble vitamin contents (expressed as g DM and normalised against fat) were significantly higher in FC (376.4 and 361.7 µg/100 g DM, for retinol and α-tocopherol, respectively) than in milk (211.4, $p < 0.05$; 84.8, $p < 0.01$) and RC (233.6, $p < 0.01$; 32.6, $p < 0.001$). In this respect, an average of 49% of the retinol and 281% of the α-tocopherol originally present in milk fat were added during the cheese-making process. Furthermore, the retinol content in milk did not differ significantly from the RC ($p > 0.05$), but the content of α-tocopherol in milk was significantly higher than in RC ($p < 0.05$; 57% lost during the cheese-making process on average). The TPC content was significantly higher in RC (54.54 mg GA equivalents/100 g DM) than in milk (18.7 mg GA equivalents, $p < 0.01$; 193% average increase over milk content during the cheese-making process) or FC (6.16 mg GA equivalents, $p < 0.001$). Moreover, the TPC content in milk was significantly higher than in FC ($p < 0.001$; 67% on average lost during the cheese-making process). The TAC was significantly higher in cheeses ($p < 0.05$) than in milk (22.4, 100.0 and 102.5 µmol Trolox equivalents/g DM, for milk, FC, and RC, respectively).

3.2. Fatty Acid Composition

Table 4; Table 5 show the FA composition in mg/g DM and in relative percentages, respectively. Saturated FAs were predominant, constituting 73.3% of FAs in milk, 74.9% in RC, and 71.1% in FC (Table 5). Monounsaturated and polyunsaturated FAs reached a value of 20.9 and 5.8% for milk, 19.2 and 5.9% for RC, and 22.2 and 6.7% for FC, respectively. The major FAs were C16:0, C18:1 n-9 cis, C18:0, and C14:0 (with percentages higher than

9.5% of total FAs). The FA composition was affected by cheese manufacturing, especially in the case of milk compared with FC and in the case of FC compared with RC.

In terms of amount, FC showed higher content in the large majority of both individual FAs and FA groups compared with the original milk (Table 4). Regarding the health indices, a higher polyunsaturated FAs (PUFA)/SFA ratio was observed in FC ($p < 0.05$, Table 4). In terms of relative percentages, the differences were observed especially in the groups of the FAs, although the individual FA C12:0 and C18:1 n-9 cis also presented significant differences ($p < 0.05$) (Table 5). FC had a lower percentage of SFA ($p < 0.05$); higher percentages of monounsaturated FAs (MUFA), PUFA, and n6 ($p < 0.05$); and a tendency toward a higher percentage of n-3 ($p < 0.10$) compared to milk (Table 5).

RC had far fewer differences in FA composition compared to the original milk than FC. In terms of amount, the RC showed a higher content of total ($p < 0.05$) and individual ($p < 0.01$) short-chain FA (SCFA), C12:0 ($p < 0.05$), and CLA *cis*-9, *trans*-11 isomer ($p < 0.05$) (Table 4). Regarding the health indices, a higher atherogenic index (AI) and a lower MUFA/SFA ratio and health-promoting index (HPI) were observed in RC ($p < 0.05$, Table 4). Similarly, the percentages of FAs were affected by the product (Table 5); the RC showed higher values of individual SCFA ($p < 0.01$), C12:0 ($p < 0.01$), C14:0 ($p < 0.01$), and CLA *cis*-9, *trans*-11 isomer ($p < 0.05$) but significantly lower percentages of C18:0 ($p < 0.001$) and total n-3.

Comparing both types of cheese, the FC showed higher content in most FAs than the RC with the exception of short and medium-chain C4-C14 FAs and some other minority FAs that were not affected by the product (Table 4). Regarding the health indices, higher MUFA/SFA, PUFA/SFA, and HPI ($p < 0.001$) and a lower AI ($p < 0.001$) and thrombogenic index (TI, $p < 0.01$) were observed in FC. Similarly, except for C16:0 and CLA *cis*-9, *trans*-11 isomer, the FA percentages were affected by the product (Table 5). FC had lower percentages of total SCFA ($p < 0.01$), C12:0 ($p < 0.001$), C14:0 ($p < 0.001$) and total SFA ($p < 0.001$) and higher percentages for the rest of the individual FAs and FA groups.

Table 3. Fat-soluble vitamins, phenolic compounds, and total antioxidant capacity of fresh and ripened Roja Mallorquina ewes' cheese and corresponding original milk.

Item [1]	Milk (M, *n* = 8)		Fresh Cheese (FC, *n* = 8)		Ripened Cheese (RC, *n* = 8)		*P* [2]		
	Mean	SEM	Mean	SEM	Mean	SEM	M vs. FC	M vs. RC	FC vs. RC
Retinol, µg/100 g DM	211.4	19.8	376.4	35.0	233.6	16.4	*	ns	**
Retinol, µg/g fat	5.56	0.45	7.89	0.77	5.45	0.58	†	ns	**
α-Tocopherol, µg/100 g DM	84.8	13.1	361.7	45.5	32.6	5.40	**	*	***
α-Tocopherol, µg/g fat	2.26	0.37	7.58	0.97	0.81	0.14	*	*	**
TPC, mg GA equivalents/100 g DM	18.7	0.54	6.16	0.55	54.5	4.44	***	**	***
TAC, µmol Trolox equivalents/g DM	22.4	6.98	100.0	12.7	102.5	11.2	*	**	ns

[1] TPC, total phenolic compounds; GA, gallic acid; TAC, total antioxidant capacity; [2] Statistical probability for comparisons: ns, not significant ($p > 0.05$); *, $p \le 0.05$; **, $p < 0.01$; ***, $p < 0.001$; †, $p \le 0.10$.

Table 4. Fatty acid (FA) content (expressed as mg/g dry matter (DM)) of fresh and ripened Roja Mallorquina ewes' cheese and corresponding original milk.

Item [1]	Milk (M, $n = 8$)		Fresh Cheese (FC, $n = 8$)		Ripened Cheese (RC, $n = 8$)		P [2]		
	Mean	SEM	Mean	SEM	Mean	SEM	M vs. FC	M vs. RC	FC vs. RC
C4:0	3.46	0.43	5.57	0.35	5.36	0.44	*	**	ns
C6:0	4.59	0.53	7.32	0.43	7.16	0.62	**	**	ns
C8:0	3.94	0.46	6.24	0.35	6.11	0.54	**	**	ns
C10:0	15.2	1.85	26.4	2.00	22.0	2.34	*	*	†
C11:0	0.29	0.03	0.45	0.05	0.38	0.03	†	ns	ns
C12:0	8.68	1.05	15.1	1.30	12.3	1.28	*	*	ns
C13:0	0.20	0.02	0.33	0.04	0.29	0.02	*	*	ns
C14:0	15.2	1.77	28.8	2.83	21.3	2.09	*	†	†
C14:1	0.62	0.07	1.25	0.15	0.87	0.08	**	*	*
C15:0	1.39	0.16	2.73	0.30	1.93	0.17	*	*	*
C15:1	0.11	0.01	0.77	0.20	0.13	0.01	**	ns	***
C16:0	44.4	5.68	81.1	10.9	49.6	4.92	*	ns	*
C16:1	2.10	0.28	3.80	0.45	2.46	0.21	*	ns	**
C17:0	0.93	0.11	1.72	0.21	1.28	0.17	*	ns	†
C17:1	0.56	0.10	1.24	0.18	0.61	0.06	*	ns	***
C18:0	21.1	2.32	37.6	5.82	19.8	2.15	†	ns	*
C18:1 n-9 *trans*	0.52	0.08	1.65	0.33	0.50	0.05	*	ns	**
C18:1 n-11 *trans* (VA)	1.98	0.22	3.47	0.51	1.79	0.17	*	ns	**
C18:1 n-9 *cis*	26.6	3.23	52.6	7.04	29.3	3.03	*	ns	*
C18:2 n-6 *trans*	0.51	0.07	1.09	0.11	0.71	0.09	*	ns	*
C18:2 n-6 *cis*	4.09	0.52	8.80	0.98	5.27	0.87	*	ns	*
γ-C18:3 n-6	0.11	0.02	0.31	0.05	0.31	0.04	*	**	ns
α-C18:3 n-3	1.08	0.14	2.60	0.38	1.24	0.15	*	ns	**
CLA *cis*-9, *trans*-11 (RA)	0.99	0.08	1.87	0.23	1.34	0.11	*	*	*
CLA *trans*-10, *cis*-12	0.06	0.01	0.25	0.03	0.09	0.01	*	ns	**
C20:0	0.35	0.04	0.17	0.03	0.12	0.01	**	**	ns
C20:1 n-9	0.14	0.01	0.14	0.01	0.11	0.02	ns	ns	ns
C20:2	0.11	0.02	0.21	0.02	0.12	0.01	*	ns	**
C20:3 n-3	0.06	0.01	0.23	0.05	0.29	0.03	**	***	ns
C20:3 n-6	0.07	0.01	0.15	0.00	0.12	0.01	**	*	†
C20:4 n-6 (ARA)	1.17	0.17	2.59	0.33	0.99	0.11	*	ns	**
C20:5 n-3 (EPA)	0.31	0.03	0.42	0.09	0.26	0.03	ns	ns	ns
C21:0	0.05	0.00	0.14	0.02	0.06	0.01	**	ns	***
C22:0	1.44	0.40	2.83	0.70	0.85	0.11	ns	ns	**
C22:1 n-9	0.10	0.01	0.16	0.01	0.13	0.02	†	†	ns
C22:2	0.06	0.01	0.08	0.01	0.03	0.00	ns	ns	*
C22:5 n-3 (DPA)	0.44	0.04	0.81	0.11	0.41	0.04	*	ns	**
C22:6 n-3 (DHA)	0.23	0.03	0.51	0.12	0.19	0.02	*	ns	**
C23:0	0.08	0.01	0.13	0.02	0.08	0.01	ns	ns	ns
C24:0	0.09	0.01	0.11	0.01	0.08	0.01	ns	ns	ns
C24:1	0.11	0.01	0.08	0.01	0.06	0.00	*	†	ns
SFA	121.5	14.1	216.9	24.9	148.7	14.7	*	ns	*
MUFA	32.9	3.91	65.2	8.75	36.0	3.51	*	ns	**
PUFA	9.32	1.03	19.9	2.37	11.4	1.31	*	ns	**
SCFA	27.2	3.10	45.6	2.99	40.6	3.86	**	*	ns
MCFA	74.5	9.16	137.3	16.4	91.2	8.97	*	ns	*
LCFA	62.0	7.15	119.1	16.7	64.3	6.74	*	ns	**
n-3	2.13	0.21	4.59	0.66	2.39	0.25	*	ns	**
n-6	5.96	0.75	12.9	1.43	7.41	0.99	*	ns	**
CLA total	1.05	0.09	2.12	0.26	1.44	0.12	*	*	*
n-6/n-3	2.76	0.13	2.91	0.14	3.08	0.17	ns	ns	ns
MUFA/SFA	0.27	0.01	0.30	0.01	0.24	0.00	ns	*	***
PUFA/SFA	0.08	0.00	0.09	0.00	0.08	0.00	*	ns	***
AI	2.71	0.08	2.52	0.08	3.11	0.02	ns	*	***
TI	3.08	0.09	2.78	0.05	3.11	0.02	ns	ns	**
HPI	0.37	0.01	0.40	0.01	0.32	0.00	ns	*	***

[1] VA, vaccenic acid; RA, rumenic acid; ARA, arachidonic acid; EPA, eicosapentaenoic acid; DPA, docosapentaenoic acid; DHA, docosahexaenoic acid; SFA, saturated FAs; MUFA, monounsaturated FAs; PUFA, polyunsaturated FAs; SCFA, short-chain FAs; MCFA, medium-chain FAs; LCFA, long-chain FAs; CLA, conjugated linoleic acid; AI (atherogenic index; (C12:0 + 4 × 14:0 + C16:0)/(MUFA + PUFA)) and TI (thrombogenic index; (C14:0 + C16:0 + C18:0)/(0.5 × MUFA + 0.5 × n-6 + 3 × n-3 + n-3/n-6)) were computed using the procedure proposed by Ulbricht and Southgate [37]; health-promoting index (HPI) was the inverse of AI, [38]; [2] Statistical probability for comparisons: ns, not significant ($p > 0.05$); *, $p \leq 0.05$; **, $p < 0.01$; ***, $p < 0.001$; †, $p \leq 0.10$.

Table 5. Fatty acid (FA) profile (expressed as % of total FA) of fresh and ripened Roja Mallorquina ewes' cheese and corresponding original milk.

Item [1]	Milk (M)		Fresh Cheese (FC)		Ripened Cheese (RC)		P [2]		
	Mean	SEM	Mean	SEM	Mean	SEM	M vs. FC	M vs. RC	FC vs. RC
C4:0-C10:0	14.7	0.85	13.7	0.68	18.3	0.22	ns	**	**
C12:0	5.21	0.12	5.03	0.19	6.20	0.09	*	**	***
C14:0	9.46	0.05	9.83	0.30	11.1	0.11	ns	**	***
C16:0	27.9	0.68	27.4	0.44	26.3	0.06	ns	†	ns
C18:0	13.5	0.21	12.7	0.41	10.6	0.22	†	***	**
C18:1 n-9 *cis*	16.9	0.55	18.00	0.21	15.6	0.08	*	ns	***
CLA *cis*-9, *trans*-11 (RA)	0.54	0.04	0.53	0.03	0.61	0.04	ns	*	ns
Others	11.8	0.23	12.8	0.30	11.3	0.05	ns	ns	**
SFA	73.3	0.61	71.1	0.35	74.9	0.09	*	ns	***
MUFA	20.9	0.53	22.2	0.35	19.2	0.05	*	†	***
PUFA	5.78	0.18	6.67	0.08	5.87	0.14	*	ns	**
n-3	1.36	0.07	1.54	0.06	1.26	0.03	†	*	**
n-6	3.74	0.13	4.44	0.05	3.88	0.17	*	ns	**

[1] SFA, saturated FAs; MUFA, monounsaturated FAs; PUFA, polyunsaturated FAs; SCFA, short-chain FAs; MCFA, medium-chain FAs; LCFA, long-chain FAs; CLA, conjugated linoleic acid; [2] Statistical probability for comparisons: ns, not significant ($p > 0.05$); *, $p \leq 0.05$; **, $p < 0.01$; ***, $p < 0.001$; †, $p \leq 0.10$.

3.3. Volatile Compounds

A total of 81 volatile compounds were detected: 17 acids, 17 alcohols, 16 ketones, 7 aromatic compounds, 8 esters, 5 aliphatic hydrocarbons, 2 sulphur compounds, 1 terpene, 3 aldehydes, and 5 lactones. Acids, alcohols, and ketones comprised almost 95% of the volatile compounds in both kinds of cheeses (FC and RC). However, acids were the most abundant family (39.7%) followed by alcohols (29.7%) in FC, while alcohols were the predominant compounds in RC (48.0%) followed by ketones (33.2%). Although both kinds of cheeses showed the same number of aromatic hydrocarbons, their concentration was higher in FC (3.0%) than in RC (1.2%). FC exhibited a greater quantity and variety of ester compounds and aliphatic hydrocarbons than RC (2.1% vs. 0.6%). Just one sulphur compound was detected in FC, but its concentration was much higher than the two detected in RC combined (0.15% vs. 0.06%). Only aldehydes and lactones were more abundant and varied in RC (0.21% and 0.31% respectively) than in FC (0.09% and 0.03%).

Acetic and 3-methylbutanoic acids were the most abundant short-chain FAs detected in FC, while hexanoic and octanoic acids predominated in RC. Ethanol was one of the predominant alcohols in both FC and RC samples. Moreover, 3-methyl-1-butan-ol and 2-butanol were present in high concentrations in FC and RC, respectively; the latter was the most concentrated volatile compound detected in RC. Butan-2-one-3-hydroxy and butan-2-one were the most abundant ketones identified in FC and RC, respectively. The former was also the predominant volatile organic compound in FC, while the latter was not present in this type of cheese. Benzene-ethanol, toluene, and *p*-xylene were the most abundant aromatic compounds that were detected in both kinds of cheese. Regarding ester compounds, ethyl acetate was the most concentrated in FC, while ethyl hexanoate had the highest concentration in RC. Lactones such as δ-octalactone and δ-decalactone were detected in both kinds of cheese, albeit in low quantities. Decanal was the only aldehyde identified in FC, while 4-heptanal was the most abundant in RC. Other compounds belonging to various minority families in this kind of product such as pentadecane, 2-ethylthio-ethanol, and farnesol were detected in both kinds of cheese.

The PCA was performed using the most discriminating volatile organic compounds present in both kinds of cheeses. The first two components (PC1 and PC2) accounted for 67.3% of the total variance, with PC1 and PC2 contributing 57.62% and 9.64%, respectively (Figure 1). FC and RC were clearly separated by PC1. Acids, alcohols, one aldehyde, and one ketone contributed positively to PC1, while alcohols, one ester, and one aromatic compound contributed negatively to PC1.

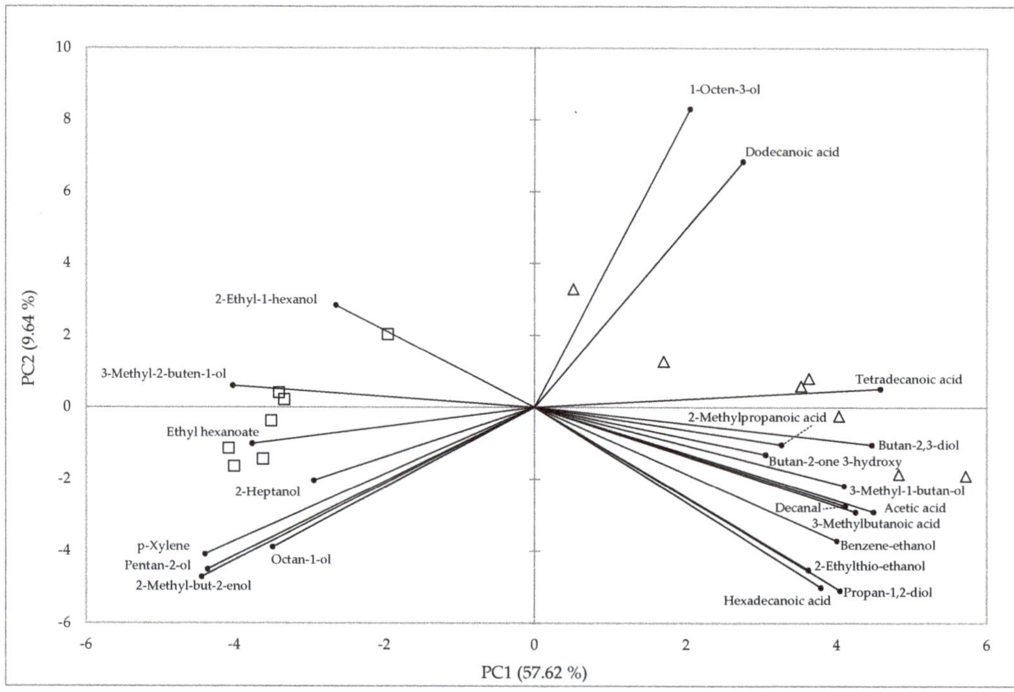

Figure 1. Principal component analysis plot representing the differentiation of Roja Mallorquina ewes' cheeses based on the main volatile compounds. △, fresh soft cheese; □, ripened cheese.

4. Discussion

4.1. Physicochemical Composition and Antioxidant Status

It is well known that several interacting factors reflect the characteristics of the original milk (breed and feeding) and its treatment (raw or pasteurised and enriched with selected starter cultures) as well as the conditions of cheese-making and ageing [17]. The breed and the ripening time are the most influential factors on the physicochemical characteristics of cheeses [25].

In general, the DM, fat, protein, and lactose contents of the milk used for the manufacture of Roja Mallorquina cheese were lower than in Guirra and Manchega breed milk [39] and higher than Assaf breed milk [40]. The main compositional characteristic of Roja Mallorquina ewes' cheese was its low fat content. These results contrast with other studies of Spanish sheep milk cheeses such as Idiazabal [41], Roncal [42], Manchego [43], and Los Pedroches [44], which found higher fat/DM content than our study. However, our results fall within the range of minimum fat/DM allowed by the European Commission's PDO regulations for Spanish sheep milk cheeses. With respect to pH, our values for ripened cheese agree with other studies [40,42–44].

Studies of fat-soluble vitamins in Spanish sheep dairy products are scarce in the scientific literature; to our knowledge, no research is available on FC. In addition, research on TPC is lacking. Regarding the vitamins, the most abundant compound in milk and RC samples was retinol, while the FC showed similar contents of retinol and α-tocopherol. Our results differ from those of Valdivielso et al. [13,45], who evaluated the fat-soluble vitamins of Spanish sheep cheeses manufactured with milk from flocks fed in grazing systems. These authors found a higher content of both vitamins, as well as the inverse pattern of α-tocopherol concentrations compared to our work (higher content in ripened cheeses than in the original milk). The vitamin contents may vary because the retinol and α-tocopherol contents of milk and cheese depend on many factors, such as the amount of herbage intake, the botanical and chemical composition of the pasture, the pasture's vegetative stage, and the concentrate feed supply in animals' diets [13,31,45].

The effect of the cheese-making process on the original milk characteristics depends on the manufacturing parameters, which thereby influence cheese features [18]. Lucas et al. [18] found that, on average, 34% of the vitamin A and 67% of the vitamin E in milk fat were lost during the cheese-making process. To a limited extent, these variations were explained by the milk fat composition, but the cheese-making process had a more substantial effect. In the present work, only the loss of α-tocopherol in the RC with respect to the original milk (over 57%) agree with these results. According to Lucas et al. [18], the loss of these micronutrients during the manufacturing process may result from their oxidative degradation by atmospheric oxygen and light as well as partial loss into the whey. In our study, the retention of these fat-soluble compounds, in particular retinol, during the cheese-making process (Table 3), could reflect the following possibilities: (a) the vitamin A of milk decreased rapidly during the first hours of light exposure, but it did not decrease further afterwards [46]; (b) casein is able to fix large quantities of retinol and α-tocopherol, and retinol that is bound to casein better resists degradation [47]; and (c) the salting process mediates the physicochemical conditions that alter the macromolecular structures of both protein and lipid molecules, and these modifications are also related to the retention of compounds in the cheese [48,49]. The increase of some fat-soluble compounds in FC, such as α-tocopherol (Table 3), may appear surprising. However, since cheese is a condensed product of milk, substances present in low concentrations in milk are more likely to be detected and even have a higher concentration in cheese [13,50]. This could reflect losses of milk proteins, lactose, fat, and minerals into whey during curd stirring and draining [51]. Finally, the type of cheese-making technology (primarily implying differences in heating, the use of starter culture, and ripening) has an important effect on the cheese composition for fat-soluble vitamins, with a higher level of retention of retinol and α-tocopherol in FC. However, compositional variability was not significantly influenced by cheese-making technology in the study by Lucas et al. [18]. Therefore, our results are not easily explained but could reflect the different processing conditions, since these affect the loss of potential cheese constituents at any stage after milking [51]. Consequently, more research is needed to elucidate the relationship between the manufacturing parameters and the fat-soluble components of cheese.

TPC variability was influenced by the cheese-making process (differences between the cheese and the original milk) and by the type of cheese-making technology (differences between cheeses). The results of this study show higher TPC concentrations in unpasteurised products (milk and RC, Table 3), which is in agreement with the results obtained by Chávez-Servín et al. [52] for goat milk, whey, and cheese. Phenolic degradation is one of the adverse effects of thermal pasteurisation that can reduce TPC concentrations [53,54]. The higher content of TPC in the RC could also be a consequence of the higher concentration of these compounds during the cheese-making process.

The TAC variability between the samples was mainly influenced by the cheese-making process, which is in agreement with the results obtained by Lucas et al. [18]. However, in this work, and unlike our results, the TAC (using the ferric reducing/antioxidant power, FRAP) of the milk was higher than the TAC of the cheese. The antioxidant capacity of natural antioxidants in dairy products is mainly due to phosphate, vitamins A and E, carotenoids, zinc, selenium, enzyme systems, oligosaccharides, and peptides [7]. They also may contain phenolic compounds [26,52,55], which have a direct effect on antioxidant activity in milk [55]. None of the antioxidant compounds analysed in this study were found to be correlated with the TAC either in milk or cheese, according to the ABTS method (data not shown). This may be because this method monitors the antioxidant capacity of both whey and total milk and is more sensitive to caseins and other low-molecular-weight compounds [56,57]. The high concentration of these proteins as a result of the loss of whey proteins and other water-soluble compounds during the cheese-making process could explain the higher TAC in cheeses than in milk.

4.2. Fatty Acid Composition

Studies involving the FA composition of Spanish sheep dairy products are limited to milk and RC, but to our knowledge, no research is available in FC. The results of FA composition in the present study are consistent with previous studies [13,45,57,58] that evaluated the FA profile of Spanish sheep cheeses manufactured in grazing systems. Milk FAs is influenced by animal feeding, and most studies on sheep grazing, as those previously mentioned, have reported that fresh pasture intake lowers the SFA content of fat, whereas that of some unsaturated FAs (UFA), such as vaccenic acid (VA), rumenic acid (RA), and n-3 FAs (ALA, α-linolenic acid; DHA, docosahexaenoic acid; EPA, eicosapentaenoic acid), increases [13,45,57,58]. With regard to the transfer of AG from milk to cheese, previous research has shown that for short (30 or fewer days) [14] or long periods of ripening (60–90 days or more) [12,13], the cheese's FA profile was similar to that of the original milk. However, in agreement with the results of the present study, the FA profile of the long ripened cheeses differed from that of the original milk [15–17]. In terms of the amount of FA by g of DM, the content of FAs was higher in cheese than in milk, especially in FC (Table 4), which is in line with the results of Valdivielso et al. [13] for Spanish RC. The authors explained that the higher concentration of FAs in cheese that maintain the same FA profile as the milk could be due to losses of main constituents of milk during the cheese-making process. This explanation may be at least partially responsible for our results, but it is clear that the different FA profile between the milk and its derived cheese must be attributed to other causes. Although comparison of these results with those of earlier studies is difficult, given the different types of cheese reported in the literature; nevertheless, it is interesting to mention the following observations.

Bergamaschi and Bittante [59] evaluated the FA profile of different dairy products from cows grazing in highland pastures, and their comparative results between fresh cheese and cheese ripened for six months are in line with those of the present study (e.g., FC had lower SFA, and AI and TI indices; higher percentages of MUFA, PUFA, n-3, and n-6). These differences, which are also detected between the FC and the original milk, could be related at least partially to the relationship between the FAs and whey proteins, as well as the effect of the processing temperature of milk during the FC production process. Indeed, and according to previous works [59,60], the link between long-chain FAs and the major protein fractions in ricotta (β-lactoglobulins) should protect the FAs against isomerisation and oxidation reactions during the high processing temperature of ricotta production. In addition, the variation of FA during manufacture and ripening could be attributed to the lipase activity present, the rennet used, and the activities of microorganisms [15,59]. The activity of these lipases is especially related to the release of free FA, which has a significant impact on the development of the characteristic flavour of cheeses. In the present study, the RC samples analysed were characterised by higher concentrations of short- and intermediate-chain FAs (C4:0–C12:0, see Table 3). Generally, these FAs have a significant impact on the development of the characteristic flavour of cheese [61] because of their low perception threshold [62]. In terms of human health, the most favourable MUFA/SFA and PUFA/SFA ratios and AI, HPI, and TI indices were observed in milk and FC, suggesting that these products may be less harmful to human health than RC [37,38,63]. However, in all the products, the n-6/n-3 ratio remained within the recommended value to prevent CVD (less than four) [64] and was similar to previous studies of sheep and goats in comparable semi-extensive systems [13,31,33,58].

Ruminant products represent the primary dietary source of RA (CLA *cis*-9, *trans*-11), which are credited with health benefits [65]. Most geometrical and positional isomers of C18:1 and C18:2, such as CLA isomers, found in milk originate from microbial hydrogenation in the rumen and subsequent enzymatic desaturation of hydrogenated intermediates in the mammary gland [66], mainly from ALA and linoleic acid (LA). Several studies have highlighted the potential of pasture grazing for enhancing the CLA proportion in milk. Thus, Valdivielso et al. [45] showed that mountain sheep milk and cheese FA profile had healthy CLA isomers (mainly RA, ≈2% of the total FAs), which were derived essentially from the intake of LA and ALA: the prevalent FAs in pasture. Furthermore, in other studies by the same authors [13,57,58], the CLA content in both milk and cheese increased progressively from indoor feeding to grazing, particularly when sheep were under mountain or part-time valley grazing. In our study, pasture intake and FA composition were not measured, but the extensive feeding regime, similar to the part-time grazing of the aforementioned studies, may be responsible for the CLA results in milk. Interestingly, cheese manufacture increased the content of RA, but the percentage of this CLA isomer was only significantly higher in the ripened cheese (Table 4; Table 5). Enhanced CLA formation at this stage could be attributed to (a) the enzymatic isomerisation of LA to CLA [67]; indeed, we observed a relationship between the percentage of LA and RA (r = −0.64, $p < 0.05$) and (b) the migration of hydrogen on linoleic acid allyl radicals to form conjugated dienyl radicals, which react with hydrogen atoms from proteins to form CLA [67,68]. This process becomes more applicable as ageing progresses, as a consequence of concurrent enzymatic hydrolysis of proteins to low-MWF (molecular-weight fractions), which are better hydrogen donors than high-MWF [15,69]. In the case of FC (made with pasteurized milk), CLA production could be affected by temperature, although the influence of this factor during the manufacture of cheeses and other dairy products is controversial [70,71]. The authors have shown that UFA, particularly with conjugated double bonds, could be sensitive to heating below 100 °C, and that these conditions could favour the formation of CLA [67,69]. In contrast, other studies on cheese manufacture [12,72,73] found that different thermal treatments did not significantly alter total CLA content or the CLA isomer profile. Therefore, future studies will be designed to further investigate the effect of the processing and storage conditions of milk and dairy products on CLA content.

4.3. Volatile Compounds

Pasteurisation affects the volatile compound profile of cheese primarily by reducing milk's indigenous microflora [74]. In our study, the milk used to manufacture FC was pasteurised, while RC samples included a starter culture. According to McSweeney [75], three sources of variability were likely involved in the volatile composition of our FC: enzymes from the rennet, indigenous milk enzymes (particularly plasmin and non-starter bacteria), and organisms that either survive pasteurisation of the milk or gain access to the pasteurised milk or curd during manufacture. In our analysis, the main families of compounds in FC were, in order of abundance: acids, alcohols, and ketones. A different sequence was observed in RC samples: alcohols, ketones, and acids. This finding agrees with Ortigosa et al. [76], who suggested that pasteurisation decreases the levels of alcohols and ketones. Juan et al. [77] also described a decrease of acids at the expense of alcohols and ketones in 15 and 60 d ewe cheese, but in this case, both kinds of cheese were pasteurized with the addition of a starter culture.

Branched-chain FAs, such as 3-methylbutanoic acid (present to a high extent in our FC samples), are derived from extensive proteolysis. Most straight-chain FAs that have between four and 20 carbon atoms, such as the hexanoic or octanoic acids detected especially in our RC samples, come from the lipolysis of triglycerides. In addition, a low proportion of FAs (e.g., acetic acid) can originate from the degradation of lactose or the oxidation of ketones, esters, or aldehydes [78].

Alcohols such as ethanol are derived primarily from the fermentation of multiple substrates such as lactose but also from amino acid metabolism and acetaldehyde reduction. Lactose fermentation can be carried out by a number of homofermentative bacteria, as well as by yeasts and leuconostocs [76]. Perhaps ethanol levels held steady in both kind of cheeses in our study, in FC because of the action of yeast (with less competition due to indigenous bacteria [79]), and in RC due to the action of the native microflora and the starter culture. Moreover, 3-methyl-1-butan-ol was also present in a high concentration in

FC. Together with 3-methylbutanoic acid, this alcoholic compound is the product of the reduction/oxidation of an intermediate aldehyde (3-methyl-butanal) undetected in our samples; 3-methyl-butanal is derived from the deamination and decarboxylation of the amino acid leucine [78].

Butan-2-one-3-hydroxy (acetoin), the main volatile compound of FC, derives from lactose and citrate metabolism and likely plays an important role in the flavour of FC [80]. Acetoin is a common compound present in many other cheeses manufactured using ewe and goat milk [74,76,81,82]. *Lactococcus lactis* ssp. *lactis* and *Lactococcus lactis* ssp. *lactisbv. diacetylactis*, which were present in the starter culture used in the manufacture of RC, are primarily responsible for the production of diacetyl [74]. The diacetyl can be reduced to acetoin, which in turn can be reduced to butan-2,3-diol, then to butan-2-one, and finally to butan-2-ol [76]. These last two compounds accounted for 50% of the volatile fraction of RC.

Benzene-ethanol, a metabolic product of yeast that is derived from phenylalanine, was detected in both cheeses as the main aromatic compound; it is responsible for floral and rose flavour notes [83]. Regarding toluene, several researchers have suggested that the compound could originate from the degradation of carotene in milk [80].

Esters play a key role in cheese aroma because of their low sensory thresholds but also due to their high volatility at room temperatures [77]. High amounts of ethanol lead to the formation of many ethyl esters [81]. Ethyl acetate and ethyl hexanoate are produced by the esterification of alcohol and acetic and hexanoic acid by microorganisms or by chemical reaction. Ethyl acetate has been positively associated with sweet odour in Appenzeller cheese, whereas ethyl hexanoate is correlated with unripe apple overtones [78].

Regarding the miscellaneous minority compounds, we can emphasise the low quantity of aldehydes (transitory compounds due to their fast reduction or oxidation to alcohols or acids), lactones (formed by the cyclisation of γ- and δ-hydroxy acids), sulphur compounds (essentially originated from methionine degradation), and aliphatic hydrocarbons (secondary products of lipid autoxidation with a minor contribution to aroma) in the samples. Finally, we can also highlight the content in both cheese samples of farnesol, which is a terpene uncommonly detected in cheese that has been identified as a non-desirable aromatic compound due to its oil and boiled vegetable aromas [84].

5. Conclusions

This is the first assessment of quality and nutritional attributes of traditional dairy products from the Roja Mallorquina sheep breed, which are linked to geographic area of origin, highlighting specific traits, such as low percentage of fat and high content of antioxidant compounds. The variability in vitamins, phenolic compounds, antioxidant capacity, and FA was influenced by the cheese-making process (differences between the cheese and the original milk) and by the type of cheese-making technology (differences between the cheeses, mainly related to heating, the use of starter culture, and ripening). Fresh soft cheese, compared to the original milk and ripened cheese, was generally characterised by better nutritional value for human health according to the fat-soluble components—a favourable level of retention of retinol and α-tocopherol and a lower saturated FA percentage and lower atherogenic and thrombogenic indices. The fresh and ripened cheeses presented different volatile profiles derived from their distinct production methods.

Given the social and economic importance of sheep's milk cheeses to small domestic producers in the Balearic Islands (Spain), this preliminary study can contribute to the recognition of the potential of the Roja Mallorquina native breed, add value to traditional products according to healthy nutritional criteria, and support the implementation of strategies to enhance commercialisation and obtain product labelling as "pasture-fed" or specific marks such as Protected Designation of Origin and Protected Geographical Indication. However, future studies are needed to further characterise the factors linked to geographical area of origin, including both natural (e.g., environment) and human (e.g., feeding management and production techniques) factors, and to investigate the relationship between the conditions of the cheese-making process and the nutritional value of cheese for human health.

Author Contributions: Conceptualisation, M.D.-P. and R.G.-P.; methodology, R.G.-P., O.P., H.G.-S., C.A., P.R.P.-C., A.H., J.L.G. and M.D.-P.; formal analysis, O.P., H.G.-S., C.A., P.R.P.-C., J.L.G. and M.D.-P.; investigation, R.G.-P.; data curation, O.P., H.G.-S., C.A., P.R.P.-C., J.L.G., A.H. and M.D.-P.; Writing—Original draft preparation, R.G.-P., O.P., H.G.-S., C.A., P.R.P.-C. and M.D.-P.; Writing—review and editing R.G.-P., O.P., H.G.-S., C.A., P.R.P.-C., A.H., J.L.G. and M.D.-P.; Supervision, project administration and funding acquisition, M.D.-P. and R.G.-P. All authors have read and agreed to the published version of the manuscript.

Funding: This research was funded by the Institute of Agricultural and Fishing Research and Training (IRFAP) of the Government of the Balearic Islands (minor contract with the Research Foundation of the University of Seville PRJ201502671-0781) and the pre-doctoral contract grant (R.G.-P.) provided by the Spanish National Institute for Agriculture and Food Research and Technology and European Social Fund (FPI2014-00013).

Institutional Review Board Statement: Not applicable.

Informed Consent Statement: Not applicable.

Data Availability Statement: The data presented in this study are available on request from the corresponding autor.

Acknowledgments: The authors express special thanks to the sheep farmers for their support and care. The authors also thank the General Service of Agricultural Research (University of Seville, Spain) and the Laboratorio Agroalimentario de Sevilla (Junta de Andalucía) for technical assistance in the laboratorial analysis.

Conflicts of Interest: The authors declare no conflict of interest. The funding sponsors had no role in the design of the study, the collection, analysis or interpretation of data, the writing of the manuscript or the decision to publish the results.

References

1. Morales-Jerrett, E.; Mancilla-Leytón, J.M.; Delgado-Pertíñez, M.; Mena, Y. The contribution of traditional meat goat farming systems to human wellbeing and its importance for the sustainability of this livestock subsector. *Sustainability* **2020**, *12*, 1181. [CrossRef]
2. ARCA. Ovella Roja Mallorquina. Available online: https://www.mapa.gob.es/es/ganaderia/temas/zootecnia/razas-ganaderas/razas/catalogo-razas/ovino/roja-mallorquina/default.aspx. (accessed on 5 March 2020). (In Spanish).
3. Vinyoles i Vidal, T.M. Notes sobre el formatge de Mallorca. *Bolletí Soc. Arqueol. Lul·Liana* **1991**, *47*, 75–88.
4. Artaud-Wild, S.M.; Connor, S.; Sexton, G.; Connor, W.E. Differences in coronary mortality can be explained by differences in cholesterol and saturated fat intakes in 40 countries but not in France and Finland. A paradox. *Circulation* **1993**, *88*, 2771–2779. [CrossRef] [PubMed]
5. Lordan, R.; Tsoupras, A.; Mitra, B.; Zabetakis, I. Dairy fats and cardiovascular disease: Do we really need to be concerned? *Foods* **2018**, *7*, 29. [CrossRef]
6. McGrath, J.; Duval, S.M.; Tamassia, L.F.M.; Kindermann, M.; Stemmler, R.T.; de Gouvea, V.N.; Acedo, T.S.; Immig, I.; Williams, S.N.; Celi, P. Nutritional strategies in ruminants: A lifetime approach. *Res. Vet. Sci.* **2018**, *116*, 28–39. [CrossRef] [PubMed]
7. Khan, I.T.; Nadeem, M.; Imran, M.; Ullah, R.; Ajmal, M.; Jaspal, M.H. Antioxidant properties of Milk and dairy products: A comprehensive review of the current knowledge. *Lipids Health Dis.* **2019**, *18*, 1–13. [CrossRef]
8. Elgersma, A.; Tamminga, S.; Ellen, G. Modifying milk composition through forage. *Anim. Feed Sci. Technol.* **2006**, *131*, 207–225. [CrossRef]
9. Capuano, E.; van der Veer, G.; Boerrigter-Eenling, R.; Elgersma, A.; Rademaker, J.; Sterian, A.; van Ruth, S.M. Verification of fresh grass feeding, pasture grazing and organic farming by cows farm milk fatty acid profile. *Food Chem.* **2014**, *164*, 234–241. [CrossRef]
10. Rubino, R. A special section on Latte Nobile: An evolving model. *J. Nutrit. Ecol. Food Res.* **2014**, *2*, 214–222. [CrossRef]
11. Méndez, C. La leche fresca de Lidl, la primera que cuenta con certificado de pastoreo y de bienestar animal. *Aral* **2018**, *1649*, 44–45.
12. Gnädig, S.; Chamba, J.F.; Perreard, E.; Chappaz, S.; Chardigny, J.M.; Rickert, R.; Steinhart, H.; Sébédio, J.L. Influence of manufacturing conditions on the conjugated linoleic acid content and the isomer composition in ripened French Emmental cheese. *J. Dairy Res.* **2004**, *71*, 367–371.

13. Valdivielso, I.; Bustamante, M.Á.; Buccioni, A.; Franci, O.; Ruíz de Gordoa, J.C.; de Renobales, M.; Barron, L.J.R. Commercial sheep flocks–fatty acid and fat-soluble antioxidant composition of milk and cheese related to changes in feeding management throughout lactation. *J. Dairy Res.* **2015**, *82*, 334–343. [CrossRef] [PubMed]

14. Bodkowski, R.; Czyż, K.; Kupczyński, R.; Patkowska-Sokoła, B.; Nowakowski, P.; Wiliczkiewicz, A. Lipid complex effect on fatty acid profile and chemical composition of cow milk and cheese. *J. Dairy Sci.* **2016**, *99*, 57–67. [CrossRef] [PubMed]

15. Laskaridis, K.; Serafeimidou, A.; Zlatanos, S.; Gylou, E.; Kontorepanidou, E.; Sagredos, A. Changes in fatty acid profile of feta cheese including conjugated linoleic acid. *J. Sci. Food Agric.* **2013**, *93*, 2130–2136. [CrossRef] [PubMed]

16. Bocquel, D.; Marquis, R.; Dromard, M.P.; Salamin, A.; Rey-Siggen, J.; Héritier, J.; Kosinska-Cagnazzo, A.; Andlauer, W. Effect of flaxseed supplementation of dairy cows' forage on physicochemical characteristic of milk and Raclette cheese. *Int. J. Dairy Technol.* **2016**, *69*, 129–136. [CrossRef]

17. Schiavon, S.; Cesaro, G.; Cecchinato, A.; Cipolat-Gotet, C.; Tagliapietra, F.; Bittante, G. The influence of dietary nitrogen reduction and conjugated linoleic acid supply to dairy cows on fatty acids in milk and their transfer to ripened cheese. *J. Dairy Sci.* **2016**, *99*, 8759–8778. [CrossRef]

18. Lucas, A.; Rock, E.; Chamba, J.F.; Verdier-Metz, I.; Brachet, P.; Coulon, J.B. Respective effects of milk composition and the cheese-making process on cheese compositional variability in components of nutritional interest. *Lait* **2006**, *86*, 21–41. [CrossRef]

19. Cifre, J.; Rigo, A.; Gulías, J.; Rallo, J.; Joy, M.; Joy, S.; Mus, M.; Sánchez, F.; Ramón, J.; Ruiz, M.; et al. *Caracterizaciò de les pastures de les Illes Balears. Quaderns dÍnvestigación 7*; Conselleria d´Agricultura i Pesca, Ed.; Govern de les Illes Balears: Palma de Mallorca, Spain, 2007.

20. AOAC. *Association of Official Analytical Chemist. Official Methods of Analysis*, 18th ed.; Horwitz, W., Latimer, G., Eds.; AOAC International: Gaithersburg, MD, USA, 2005.

21. Delgado-Pertíñez, M.; Alcalde, M.J.; Guzmán-Guerrero, J.L.; Castel, J.M.; Mena, Y.; Caravaca, F. Effect of hygiene-sanitary management on goat milk quality in semi-extensive systems in Spain. *Small Rumin. Res.* **2003**, *47*, 51–61. [CrossRef]

22. ISO/IDF. *Cheese—Determination of Fat Content—Van Gulik Method*; ISO 3433:2008 (IDF 222:2008); ISO: Genève, Switzerland, 2008.

23. AOAC. *Association of Official Analytical Chemists. Official Methods of Analysis*, 16th ed.; Cunniff, P., Ed.; AOAC International: Washington, DC, USA, 1999.

24. AOAC. *Association of Official Analytical Chemists. Official Methods of Analysis*, 15th ed.; Helrich, K., Ed.; AOAC International: Arlington, VA, USA, 1990.

25. De la Haba Ruiz, M.; Ruiz Pérez-Cacho, P.; Dios Palomares, R.; Galan-Soldevilla, H. Classification of artisanal Andalusian cheeses on physicochemical parameters applying multivariate statistical techniques. *Dairy Sci. Technol.* **2016**, *96*, 95–106. [CrossRef]

26. Guzmán, J.L.; Pérez-Écija, A.; Zarazaga, L.A.; Martín-García, A.I.; Horcada, A.; Delgado-Pertíñez, M. Using dried orange pulp in the diet of dairy goats: Effects on milk yield and composition and blood parameters of dams and growth performance and carcass quality of kids. *Animal* **2020**, *14*, 2212–2220. [CrossRef]

27. Sukhija, P.S.; Palmquist, D.L. Rapid method for determination of total fatty acid content and composition of feedstuffs and feces. *J. Agri. Food Chem.* **1988**, *36*, 1202–1206. [CrossRef]

28. Juárez, M.; Polvillo, O.; Contò, M.; Ficco, A.; Ballico, S.; Failla, S. Comparison of four extraction/methylation analytical methods to measure fatty acid composition by gas chromatography in meat. *J. Chromatogr. A* **2008**, *1190*, 327–332. [CrossRef] [PubMed]

29. Herrero-Barbudo, M.C.; Granado-Lorencio, F.; Blanco-Navarro, I.; Olmedilla-Alonso, B. Retinol, α-and γ-tocopherol and carotenoids in natural and vitamin A- and E-fortified dairy products commercialized in Spain. *Int. Dairy J.* **2005**, *15*, 521–526. [CrossRef]

30. Chauveau-Duriot, B.; Doreau, M.; Nozière, P.; Graulet, B. Simultaneous quantification of carotenoids, retinol, and tocopherols in forages, bovine plasma, and milk: Validation of a novel UPLC method. *Anal. Bioanal. Chem.* **2010**, *397*, 777–790. [CrossRef]

31. Gutiérrez-Peña, R.; Fernández-Cabanás, V.M.; Mena, Y.; Delgado-Pertíñez, M. Fatty acid profile and vitamins A and E contents of milk in goat farms under Mediterranean wood pastures as affected by grazing conditions and seasons. *J. Food Compost. Anal.* **2018**, *72*, 122–131. [CrossRef]

32. Pellegrini, N.; Ke, R.; Yang, M.; Rice-Evans, C. Screening of dietary carotenoids and carotenoid-rich fruit extracts for antioxidant activities applying 2,2 0-azinobis(3-ethylenebenzothiazoline-6-sulfonic acid) radical catión decolourisation assay. *Meth. Enzymol.* **1999**, *299*, 379–389.

33. Delgado-Pertíñez, M.; Gutiérrez-Peña, R.; Mena, Y.; Fernández-Cabanás, V.M.; Laberye, D. Milk production, fatty acid composition and vitamin E content of Payoya goats according to grazing level in summer on Mediterranean shrublands. *Small Rumin. Res.* **2013**, *114*, 167–175.

34. Revilla, I.; González-Martín, M.I.; Vivar-Quintana, A.M.; Blanco-López, M.A.; Lobos-Ortega, I.A.; Hernández-Hierro, J.M. Antioxidant capacity of different cheeses: Affecting factors and prediction by near infrared spectroscopy. *J. Dairy Sci.* **2016**, *99*, 5074–5082. [CrossRef]

35. Vázquez, C.V.; Rojas, M.G.; Ramirez, C.A.; Chávez-Servin, J.L.; García-Gasca, T.; Ferriz-Martínez, R.A.; Garcia, O.P.; Rosado, J.L.; López-Sabater, C.M.; Castellote, A.I.; et al. Total phenolic compounds in milk from different species. Design of an extraction technique for quantification using the Folin-Ciocalteu method. *Food Chem.* **2015**, *176*, 480–486. [CrossRef]

36. Guzmán, J.L.; Delgado Pertíñez, M.; Galán Soldevilla, H.; Ruiz Pérez-Cacho, P.; Polvillo Polo, O.; Zarazaga, L.Á.; Avilés Ramírez, C. Effect of citrus by-product on physicochemical parameters, sensory analysis and volatile composition of different kinds of cheese from raw goat milk. *Foods* **2020**, *9*, 1420.

37. Ulbricht, T.L.V.; Southgate, D.A.T. Coronary heart disease: Seven dietary factors. *Lancet* **1991**, *338*, 985–992. [CrossRef]
38. Chen, S.; Bobe, G.; Zimmerman, S.; Hammond, E.G.; Luhman, C.M.; Boylston, T.D.; Freeman, A.E.; Beitz, D.C. Physical and sensory properties of dairy products from cows with various milk fatty acid compositions. *J. Agric. Food Chem.* **2004**, *52*, 3422–3428. [CrossRef] [PubMed]
39. Jaramillo, D.P.; Zamora, A.; Guamisa, B.; Rodríguez, M.; Trujillo, A.J. Cheesemaking aptitude of two Spanish dairy ewe breeds: Changes during lactation and relationship between physico-chemical and technological properties. *Small Rumin. Res.* **2008**, *78*, 48–55. [CrossRef]
40. Estrada, O.; Ariño, A.; Juan, T. Salt Distribution in Raw Sheep Milk Cheese during Ripening and the Effect on Proteolysis and Lipolysis. *Foods* **2019**, *8*, 100. [CrossRef]
41. Hernández, I.; Barrón, L.J.R.; Virto, M.; Pérez-Elortondo, F.J.; Flanagan, C.; Rozas, U.; Nájera, A.I.; Albisu, M.; Vicente, M.S.; de Renobales, M. Lipolysis, proteolysis and sensory properties of ewe's raw milk cheese (Idiazabal) made with lipase addition. *Food Chem.* **2009**, *116*, 158–166. [CrossRef]
42. Irigoyen, A.; Izco, J.M.; Ibáñez, F.C.; Torre, P. Influence of calf or lamb rennet on the physicochemical, proteolytic, and sensory characteristics of an ewes milk cheese. *Int. Dairy J.* **2002**, *12*, 27–34. [CrossRef]
43. Cabezas, L.; Sánchez, I.; Poveda, J.M.; Seseña, S.; Palop, M.L.L. Comparison of microflora, chemical and sensory characteristics of artisanal Manchego cheeses from two dairies. *Food Control* **2007**, *18*, 11–17. [CrossRef]
44. San Juan, E.; Millan, R.; Saavedra, P.; Carmona, M.A.; Gómez, R.; Fernandez-Salguero, J. Influence of animal and vegetable rennet on the physicochemical characteristics of Los Pedroches cheese during ripening. *Food Chem.* **2002**, *78*, 281–289. [CrossRef]
45. Valdivielso, I.; Bustamante, M.A.; Aldezabal, A.; Amores, G.; Virto, M.; Ruíz de Gordoa, J.C.; de Renobales, M.; Barron, L.J.R. Case study of a commercial sheep flock under extensive mountain grazing: Pasture derived lipid compounds in milk and cheese. *Food Chem.* **2016**, *197*, 622–633. [CrossRef]
46. De Man, J.M. Light-induced destruction of vitamin A in milk. *J. Dairy Sci.* **1981**, *64*, 2031–2032. [CrossRef]
47. Poiffait, A.; Lietaer, E.; Le Pavec, P.; Adrian, J. Interrelations physico-chimiques et nutritionnelles entre la caséine et les vitamines liposolubles. *Ind. Aliment. Agric.* **1992**, *109*, 939–947.
48. Bansal, V.; Mishra, S.K. Reduced-sodium cheeses: Implications of reducing sodium chloride on cheese quality and safety. *Compr. Rev. Food Sci. Food Saf.* **2020**, *19*, 733–758. [CrossRef] [PubMed]
49. do Oriente, S.F.; Barreto, F.; Tomaszewski, C.A.; Barnet, L.S.; Souza, N.C.; Oliveira, H.M.; de Bittencourt, M.A. Retention of vitamin A after goat milk processing into cheese: A nutritional strategy. *J. Food Sci. Technol.* **2020**, *57*, 4364–4370. [CrossRef] [PubMed]
50. Poulopoulou, I.; Zoidis, E.; Massouras, T.; Hadjigeorgiou, I. Terpenes transfer to milk and cheese after oral administration to sheep fed indoors. *J. Anim. Physiol. Anim. Nutr.* **2012**, *96*, 172–181. [CrossRef]
51. Abd El-Gawad, M.A.M.; Ahmed, N.S. Cheese yield as affected by some parameters. Review. *Acta Sci. Pol. Technol. Aliment.* **2011**, *10*, 131–153.
52. Chávez-Servín, J.L.; Andrade-Montemayor, H.M.; Velázquez Vázquez, C.; Aguilera Barreyro, A.; García-Gasca, T.; Ferríz Martínez, R.A.; Olvera Ramírez, A.M.; de la Torre-Carbot, K. Effects of feeding system, heat treatment and season on phenolic compounds and antioxidant capacity in goat milk, whey and cheese. *Small Rumin. Res.* **2018**, *160*, 54–58. [CrossRef]
53. Nagy, S.; Rouseff, R.; Lee, H. Thermally degraded flavors in citrus juice products. *ACS Symp. Ser.* **1989**, *409*, 331–345.
54. Chen, Y.; Yu, L.J.; Rupasinghe, H.P. Effect of thermal and non-thermal pasteurization on the microbial inactivation and phenolic degradation in fruit juice: A mini-review. *J. Sci. Food Agric.* **2013**, *93*, 981–986. [CrossRef]
55. Di, T.A.; Bonanno, A.; Cecchini, S.; Giorgio, D.; Di, G.A.; Claps, S. Effects of Sulla forage (*Sulla coronarium* L.) on the oxidative status and milk polyphenol content in goats. *J. Dairy Sci.* **2015**, *98*, 37–46.
56. Zulueta, A.; Maurizi, A.; Frígola, A.; Esteve, M.J.; Coli, R.; Burini, G. Antioxidant capacity of cow milk, whey and deproteinized milk. *Int. Dairy J.* **2009**, *19*, 380–385. [CrossRef]
57. Virto, M.; Bustamante, M.; de Gordoa, J.C.R.; Amores, G.; Fernández-Caballero, P.N.; Mandaluniz, N.; Arranz, J.; Nájera, A.I.; Albisu, M.; Pérez-Elortondo, F.J.; et al. Interannual and geographical reproducibility of the nutritional quality of milk fat from commercial grazing flocks. *J. Dairy Res.* **2012**, *79*, 485–494. [CrossRef] [PubMed]
58. Abilleira, E.; Collomb, M.; Schlichtherle-Cerny, H.; Virto, M.; de Renobales, M.; Barron, L.J.R. Winter/Spring changes in fatty acid composition of farmhouse Idiazabal cheese due to different flock management systems. *J. Agric. Food Chem.* **2009**, *57*, 4746–4753. [CrossRef] [PubMed]
59. Bergamaschi, M.; Bittante, G. Detailed fatty acid profile of milk, cheese, ricotta and by products, from cows grazing summer highland pastures. *J. Dairy Res.* **2017**, *84*, 329–338. [CrossRef] [PubMed]
60. Pérez, M.D.; Calvo, M. Interaction of β-lactoglobulin with retinol and fatty acids and its role as a possible biological function for this protein: A review. *J. Dairy Sci.* **1995**, *78*, 978–988. [CrossRef]
61. Martino, C.; Ianni, A.; Grotta, L.; Pomilio, F.; Martino, G. Influence of Zinc Feeding on Nutritional Quality, Oxidative Stability and Volatile Profile of Fresh and Ripened Ewes' Milk Cheese. *Foods* **2019**, *8*, 656. [CrossRef]
62. Molimard, P.; Spinnler, H.E. Review: Compounds involved in the flavour of surface mould-ripened cheeses: Origins and properties. *J. Dairy Sci.* **1996**, *79*, 169–184. [CrossRef]
63. Gil, A.; Martínez de Victoria, E.; Olza, J. Indicators for the evaluation of diet quality. *Nutr. Hosp.* **2015**, *31*, 128–144.

64. Molendi-Coste, O.; Legry, V.; Leclercq, I.A. Why and how meet n-3 PUFA dietary recommendations? A review. *Gastroenterol. Res. Pract.* **2011**, *2011*, 364040. [CrossRef]
65. MacRae, J.; O'Reilly, L.; Morgan, P. Desirable characteristics of animal products from a human health perspective. *Livest. Prod. Sci.* **2005**, *94*, 95–103. [CrossRef]
66. Palmquist, D.L.; Lock, A.L.; Shingfield, K.J.; Bauman, D.E. Biosynthesis of conjugated linoleic acid in ruminants and humans. *Adv. Food Nutr. Res.* **2005**, *50*, 179–217.
67. Ha, Y.L.; Grimm, N.K.; Pariza, M.W. Newly recognized anticarcinogenic fatty acids: Identification and quantification in natural and processed cheeses. *J. Agric. Food Chem.* **1989**, *37*, 75–81. [CrossRef]
68. Shantha, N.C.; Decker, E.A. Conjugated linoleic acid concentrations in processed cheese containing hydrogen donors, iron and dairy based additives. *Food Chem.* **1993**, *47*, 257–261. [CrossRef]
69. Shantha, N.C.; Decker, E.A.; Ustunol, Z. Conjugated linoleic acid concentration in processed cheese. *J. Am. Oil Chem. Soc.* **1992**, *69*, 425–428. [CrossRef]
70. Buccioni, A.; Rapaccini, S.; Antongiovanni, M.; Minieri, S.; Conte, G.; Mele, M. Conjugated linoleic acid and C18:1 isomers content in milk fat of sheep and their transfer to Pecorino Toscano cheese. *Int. Dairy J.* **2010**, *20*, 190–194. [CrossRef]
71. Avilez, J.P.; Wladimir, M.; Delgado-Pertiñez, M. Conjugated linoleic acid of dairy foods is affected by cow's feeding system and processing of milk. *Sci. Agric.* **2016**, *73*, 103–108.
72. Luna, P.; De la Fuente, M.A.; Juárez, M. Conjugated linoleic acid in processed cheeses during the manufacturing stages. *J. Agric. Food Chem.* **2005**, *53*, 2690–2695. [CrossRef]
73. Luna, P.; Juárez, M.; De la Fuente, M.A. Conjugated linoleic acid content and isomer distribution during ripening in three varieties of cheeses protected with designation of origin. *Food Chem.* **2007**, *103*, 1465–1472. [CrossRef]
74. Buchin, S.; Delague, V.; Duboz, G.; Berdague, J.L.; Beuvier, E.; Pochet, S.; Grappin, R. Influence of pasteurization and fat composition of milk on the volatile compounds and flavor characteristics of a semi-hard cheese. *J. Dairy Sci.* **1998**, *81*, 3097–3108. [CrossRef]
75. McSweeney, P.; Sousa-Gallagher, M.J. Biochemical pathways for the production of flavour compounds in cheeses during ripening: A review. *Lait* **2000**, *80*, 293–324. [CrossRef]
76. Ortigosa, M.; Torre, P.; Izco, J.M. Effect of pasteurization of ewe's milk and use of a native starter culture on the volatile components and sensory characteristics of Roncal cheese. *J. Dairy Sci.* **2001**, *84*, 1320–1330. [CrossRef]
77. Juan, B.; Barron, L.J.R.; Ferragut, V.; Trujillo, A.J. Effects of high pressure treatment on volatile profile during ripening of ewe milk cheese. *J. Dairy Sci.* **2007**, *90*, 124–135. [CrossRef]
78. Curioni, P.M.G.; Bosset, J.O. Key odorants in various cheese types as determined by gas chromatography-olfactometry. *Int. Dairy J.* **2002**, *12*, 959–984. [CrossRef]
79. Olarte, C. Caracterización del Queso de Cameros. Evolución de Parámetros Fisicoquímicos y Microbiológicos Durante su Maduración. Ph.D. Thesis, Univ. Rioja, La Rioja, Spain, 1999.
80. Atasoy, A.F.; Hayaloglu, A.A.; Kırmacı, H.; Levent, O.; Türkoğlu, H. Effects of partial substitution of caprine for ovine milk on the volatile compounds of fresh and mature Urfa cheeses. *Small Rumin. Res.* **2013**, *115*, 113–123. [CrossRef]
81. Karali, F.; Georgala, A.; Massouras, T.; Kaminarides, S. Volatile compounds and lipolysis levels of Kopanisti, a traditional Greek rawmilk cheese. *J. Sci. Food Agric.* **2013**, *93*, 1845–1851. [CrossRef]
82. Quintanilla, P.; Hettinga, K.A.; Beltrán, M.C.; Escriche, I.; Molina, M.P. Short communication: Volatile profile of matured Tronchón cheese affected by oxytetracycline in raw goat milk. *J. Dairy Sci.* **2020**, *103*, 6015–6021. [CrossRef]
83. Güler, Z. Profiles of organic acid and volatile compounds in acid-type cheeses containing herbs and spices (Surk cheese). *Int. J. Food Prop.* **2014**, *17*, 1379–1392. [CrossRef]
84. Risner, D.; Tomasino, E.; Hughes, P.; Meunier-Goddik, L. Volatile aroma composition of distillates produced from fermented sweet and acid whey. *J. Dairy Sci.* **2019**, *102*, 202–210. [CrossRef]

Article

Effect of Citrus By-product on Physicochemical Parameters, Sensory Analysis and Volatile Composition of Different Kinds of Cheese from Raw Goat Milk

José Luis Guzmán [1], Manuel Delgado Pertíñez [2], Hortensia Galán Soldevilla [3], Pilar Ruiz Pérez-Cacho [3], Oliva Polvillo Polo [4], Luis Ángel Zarazaga [1] and Carmen Avilés Ramírez [3,*]

[1] Departamento de Ciencias Agroforestales, Escuela Técnica Superior de Ingeniería, Universidad de Huelva, "Campus de Excelencia Internacional Agroalimentario, ceiA3", Campus de la Rábida, Palos de la Frontera, 21819 Huelva, Spain; guzman@uhu.es (J.L.G.); zarazaga@uhu.es (L.Á.Z.)

[2] Departamento de Ciencias Agroforestales, ETSIA, Universidad de Sevilla, 41013 Sevilla, Spain; pertinez@us.es

[3] Departamento de Bromatología y Tecnología de los Alimentos, Universidad de Córdoba, Campus de Rabanales, 14070 Córdoba, Spain; bt1gasoh@uco.es (H.G.S.); pilar.ruiz@uco.es (P.R.P.-C.)

[4] Centro de Investigación, Tecnología e Innovación, Universidad de Sevilla, Avda. Reina Mercedes 4-B, 41012 Sevilla, Spain; oppolo@us.es

* Correspondence: v92avrac@uco.es; Tel.: +34-957-218526

Received: 7 September 2020; Accepted: 4 October 2020; Published: 8 October 2020

Abstract: The increased use of concentrates to reduce pasture as a feed source in productive systems like Payoya breed goat farms has made it necessary to decrease feeding costs. The inclusion of agro-industry by-products such as dry orange pulp pellets in goat diets has been suggested as a sustainable alternative to cereal-based concentrates. The aim of this work was to assess the influence of diets including dry orange pulp pellets on the quality of cheeses traditionally made from Payoya breed goat milk. We analysed the physicochemical characteristics, sensory properties and volatile compound profiles of 18 artisanal cheeses made from raw Payoya milk. In this study, goats were fed with different concentrations of dry orange pulp; and cheeses were curdled with animal and vegetable coagulants. Slight differences were detected between some cheeses. However, the use of citrus by-products in the Payoya goat diets did not substantially affect the cheeses' physicochemical properties, olfactory attributes, or volatile profiles. Therefore, dried citrus pulp can be used as a substitute for cereal concentrates without affecting the distinct properties of these ripened raw goat milk cheeses.

Keywords: goat cheese; odour; raw milk; volatile compounds

---

1. Introduction

The demanding European legislation on food quality and security issues and the increase of practices aiming to obtain growing dairy yields per farm led to the intensification of production and the decline of grazing livestock systems [1]. This dairy intensification has led to an increase in the use of concentrates and reducing or eliminating pasture, as in Payoya breed farms [2]. This breed is a goat population traditionally reared under extensive or semi-extensive production systems, well-adapted to the low winter and high summer temperatures, and prevailing in the regions of southern Spain where they are most abundant [3]. As a result of this intensification, it is necessary to decrease livestock feeding costs by developing strategies, such as greater dependence on local feed resources, to increase the sustainability of livestock production systems [4,5].

Most cheese production from Spanish goat milk is performed by large dairy operations. However, small local industries and artisanal farm dairies still play a role in the industry, providing added value with their high-quality products. The cheese from Payoya goats milk is an artisanal product made of raw milk, usually curded with animal rennet without using starter cultures. These kinds of cheeses are ripened over different periods of time depending on the final product and are produced in the mountains of Grazalema in Cadiz (Spain) and in its surroundings. The shape of the cheese is cylindrical (20–35 cm diameter × 10–15 cm height) and its weight ranges from 1.5 to 3.5 kg. Its crust is hard, oily, and slightly unctuous and its paste is compact and greasy with small eyes irregularly distributed [6].

The use of alternative milk coagulant enzymes is an interesting investigation subject for making cheese [7]. Vegetable enzymes have been widely investigated as possible coagulants in cheese manufacture [7], however, some of them have been found to be inappropriate for cheese production due to a characteristic excessive proteolytic activity that drops cheese yield and produces un-desirable flavors in the final product. The cardoon flower (*Cynara cardunculus*) deserves special mention among the vegetable enzymes because it produces acceptable final products. This coagulant is traditionally used as an alternative to animal rennet in the manufacture of different Spanish and Portuguese artisanal ewe- and/or goat-milk cheeses. The higher values of soluble nitrogen and the lower content of residual casein are responsible for the pronounced and pleasant taste of ewe and goat cheeses made with this vegetable coagulant [7].

Small local producers play a key role in the sustainable rural development of the areas where they are located, contributing to the conservation of zones of high ecological value like the mountain range of Cádiz and Málaga. Other strategic priorities for the development of the goat sector in Spain include diversifying dairy products to increase milk demand and improving the competitiveness of current production systems by using by-products of the agri-food industry [8].

Spain produces more than 3.6 million tonnes of oranges per year. In 2018, Spain was the primary producer of the European Union and the sixth-largest global producer [9]. The principal citrus by-product, orange pulp, can partially replace cereal grains in ruminant feedstuffs with no adverse effect on milk yield or composition, and may even improve the sensory characteristics of milk-based products such as cheese [10]. The inclusion of dry orange pulp in concentrates as a substitute for traditionally used cereals constitutes a sustainable and more effective way of using this by-product. However, few studies have addressed the by-product's effect on the sensory properties of goat cheeses [10], and none have examined the cheeses' volatile compounds. Moreover, to our knowledge, this is the first time that cheeses made with the native Payoya breed have been studied. Despite these cheeses having been greatly appreciated for their organoleptic quality which has been recognized by international organisations (seven awards at "2019 World Cheese Awards" in Bergamo, Italy), their aroma and volatile compounds have never been well characterized.

Thus, the aim of this work was to assess the influence of dietary dry orange pulp pellets on the physicochemical characteristics, sensory properties, and volatile compound profiles of cheeses traditionally made from Payoya milk and curdled with animal and vegetable rennet.

2. Materials and Methods

2.1. Animals and Experimental Rations

This study was performed at the experimental farm of the University of Huelva (Huelva, Spain). Forty-four primiparous Payoya breed goats were allocated to three experimental groups, each of them with a different diet. The three experimental diets were the following: control (C; $n = 14$), with a commercial concentrate and alfalfa hay as forage; diet 1 (DOP40; $n = 16$) based on C but with 40% of the cereals in the concentrate substituted with dried orange pulp (DOP); and diet 2 (DOP80; $n = 14$), based on C, but with 80% of the cereals in the concentrate substituted with DOP. DOP pellets were prepared using orange juice residues, following a conventional industrial process (Cítricos del Andévalo, S.A., Huelva, Spain). Briefly, the residue (pulp) of the orange obtained after the extraction of the juice was pressed to reduce the humidity to 70%. Afterwards, and after crushing and adding calcium oxide to facilitate its drying, this pressed pulp was dried in a rotary drum dryer until it reached a humidity of 10%. Finally, before proceeding to pelletization, the liquid extracted after the pressing was dehydrated and incorporated into the dried pulp.

In the fifth month of lactation, the animals were offered the experimental diets adapted to this lactation month. The formulation of the rations was designed using the Feed Ration Balancer (Format Solutions) software, version 2.0 (2017; Cargill, Inc., Minneapolis, MN USA; www.formatsolutions.com). The chemical composition of the isoenergetic and isoproteic diets are described in Table 1. Food intake for each group was calculated daily, in late lactation (120–180 d), by subtracting the orts (uneaten food) from the amount of food offered every day. The total average dry matter (DM) intake per goat in the diet groups was 1.78, 1.76, and 1.75 kg/day in the C, DOP40, and DOP80 groups, respectively. For more details on animal management until early lactation, see Guzmán et al. [5].

Table 1. Ration ingredients, proximate composition, and nutritive value of the experimental diets used to feed goats during the fifth month of lactation.

Ration Ingredients, % DM	Lactation Experimental Diets [1]		
	Control	DOP40	DOP80
Alfalfa hay	20.16	20.28	20.44
Concentrate			
Dehydrated orange pulp (pellets)	0.00	19.36	38.64
Grain oats	21.44	12.83	4.24
Grain barley	8.28	4.96	1.65
Grain corn	18.76	11.25	3.77
Soy flour, 44%	7.09	9.92	12.57
Sunflower pellets, 28%	12.46	12.12	13.35
Grain peas	10.01	7.87	3.93
Salt	0.39	0.39	0.39
Stabilised lard	0.39	0.00	0.00
Vitamins and minerals [2]	1.01	1.01	1.02
Proximate Composition and Nutritive Value, % DM			
DM, %	87.08	87.08	88.09
Crude protein	20.92	18.66	18.30
Neutral detergent fibre	29.82	26.56	28.29
Acid detergent fibre	14.69	15.24	16.83
Acid detergent lignin	3.09	3.13	3.43
Sugar and starch	36.07	36.07	20.49
Ether extract	2.63	1.85	1.43
Ash	6.50	7.47	8.64
Calcium	0.60	0.96	1.27
Phosphorus	0.48	0.41	0.39
Gross energy, kcal/g DM	4.37	4.31	4.25
Forage unit for lactation, UFL/kg	0.98	0.98	0.96
Protein digestible in the intestine (PDI)	10.42	10.42	11.42

[1] Control, diet based on commercial concentrates plus alfalfa hay; DOP40, diet based on concentrate with 40% of cereals replaced by DOP plus alfalfa hay; DOP80, diet based on concentrate with 80% of cereals replaced by DOP plus alfalfa hay. [2] Nutral cabras LD granulado, Cargill®, Spain. DM: dry matter.

2.2. Cheese Manufacture and Sampling

Eighteen cheeses were manufactured for this study, including three replicate samples for each diet group (C, DOP40, and DOP80) and rennet type (animal and vegetable). In the fifth month of lactation, about 20 kg of bulk milk per batch was collected from each experimental ration group and was transported in a refrigerated vehicle to an artisanal factory for cheese manufacture. Another two batches were produced in two consecutive days, but under the same conditions. Half of each batch was clotted using a commercial animal rennet and the other half, using a commercial vegetable rennet, according to the manufacturer's instructions. The cheeses were made with raw milk (in the fifth month of lactation) and without adding a starter culture, following traditional methods [11]. Briefly, after heating to 32 °C, the animal (Avances Bioquímicos Alimentación S.L., Pontevedra, Spain; about 0.25 mL per litre of milk) or vegetal (thistle *Cynara cardunculus* L., Avances Bioquímicos Alimentación S.L.; about 0.25 mL per litre of milk) rennet was added to obtain clotting in 60 min. After coagulation, the curd was cut with a lyre of parallel wires to obtain grains the size of a hazelnut. Then, the temperature was raised to 34 °C and the curd was stirred mechanically for approximately 35 min, before whey drainage. The curds were moulded into pieces and pressed in a hydraulic press (2.0 kg/cm^2) for 1 h. Finally, the cheeses were immersed in brine (15–18 °Baumé, 6 °C, pH = 5.15–5.20) for 30 min. Afterwards, the cheeses were ripened in chambers at 10–12 °C with a relative humidity of 85% for 60 days.

2.3. Physico-Chemical Analysis

Total solids (TS, g/100 g cheese), pH, fat (g/100 g cheese), fat/TS (g/100 g TS), and sodium chloride (g/100 g cheese) were analysed according to De la Haba et al. [11]. Fat content was measured according to the FIL-IDF methods [12]. TS content was determined following the official method [13]. The pH was measured with a pH metre (HANNA FHT-803) with a pH electrode. The sodium chloride content was analysed using back titration with potassium thiocyanate to determine the concentration of chloride ions in the solution based on the Volhard method [13]. All determinations were made in duplicate and each pair of data was averaged.

2.4. Sensory Analysis

2.4.1. Sample Preparation

The samples were prepared according to Ruiz Pérez-Cacho et al. [14]. Each taster received one portion of cheese per sample. Three to four samples were served, one at a time, over a session. Mineral water was used to cleanse the palate between samples.

2.4.2. Assessors

Eight highly trained panellists from the Sensory Laboratory at the University of Córdoba (Spain) collaborated in this research. The panel was selected and trained following the ISO [15–17]. These assessors had previous experience in the sensory analysis of several foods [18–24] and had undergone specific training in cheeses [14]. Testing was performed at the sensory test area under the conditions specified in the ISO [25]. All analyses were conducted in the morning.

2.4.3. Sensory Profile

The methodology followed is based on the ISO [16,26,27]. The odour profile was made following Ruiz Pérez-Cacho et al. [14]. Ten odour attributes were analysed on a non-structured scale of 10 cm (overall intensity, milk, butter, heated milk, cake, toffee, nuts, goat and butyric/propionic acid). All evaluations were made in duplicate.

2.5. Volatile Compounds

Volatile compounds were extracted by headspace solid-phase microextraction (SPME) from 5 g of minced and homogenised cheese samples. The samples were deposited in 50 mL vials and heated while stirring at 40 °C. A fibre of divinylbenzene/carboxen/polydimethylsiloxane (DVB-CAR-PDMS; 1 cm long × 110 μm diameter; Supelco, Bellefonte, PA, USA) was fixed in the headspace of the vial for 10 min. The volatile compounds were desorbed into the split-splitless injector of the gas chromatograph (GC) system set at 250 °C for 5 min. The volatile compounds were analysed in a GC Thermo–Scientific Trace 1300 connected to an ISQ mass spectrometer (MS) using a VF-42

WAXms column (30 m × 250 µm i.d. × 0.50 µm film thickness) with helium as the carrier gas. The chromatographic conditions were as follows: the oven temperature began at 45 °C for 4 min, increased to 150 °C at 5 °C/min, remained at 150 °C for 3 min, increased to 250 °C at 6 °C/min, and remained at 250 °C for 5 min; the transfer line temperature was 280 °C. The MS worked in electron impact mode. The electron impact energy was 70 eV and the equipment recorded data at a rate of 1 scan/s. The relative abundance of the volatile compounds in the chromatograms was calculated by considering the area units under each peak. A series of n-alkanes was used to obtain retention index (RI) values for each volatile compound under the same conditions. Compounds were tentatively identified by comparing their mass spectra with those contained in the National Institute of Standards and Technology (NIST; Gaithersburg, MD, USA) library or in previously published literature.

2.6. Statistical Analysis

All statistical tests were performed with the IBM SPSS Statistics for Windows (version 26.0; IBM Corp., Armonk, NY, USA). A basic descriptive statistical analysis (mean and standard deviation) and a two-way ANOVA (rennet × feeding) were applied for each physicochemical parameter and sensory attribute, followed by Tukey test ($p < 0.05$). In addition, a one-way ANOVA was applied for each sensory attribute to test mean differences between assessors. Finally, a multivariate analysis was performed with the principal component analysis (PCA) command of the XLSTAT software (Addinsoft Inc., New York, NY, USA). We carried out a principal component analysis using a Pearson correlation matrix on the mean values for descriptive measures of the sensory analysis and volatile compound content.

3. Results and Discussion

3.1. Physicochemical Analysis

Table 2 presents the means, standard deviations, and ANOVA (rennet × feeding) of physicochemical parameters (F and probability values). The effect of rennet was significant for all physicochemical parameters ($p < 0.01$) except for pH. The animal coagulant showed the highest average TS and fat content, and the lowest salt level. The pH, fat content, fat/TS value, and sodium chloride level were affected by the type of diet used ($p < 0.05$). Cheeses from goats fed with a diet based on DOP pellets had higher average pH and salt levels, and lower fat content than the cheeses from goats fed with the control diet. Finally, there was a rennet × feeding interaction effect for all the parameters studied ($p < 0.05$, Table 2). Compared to other studies, we found slightly lower values than other Spanish goat milk cheeses [11,28–34].

Table 2. Descriptive measures (mean and standard deviation) and analysis of variance (rennet × diet) of physicochemical parameters (F and probability values).

	Effect	pH	TS (g/100 g Cheese)	Fat (g/100 g Cheese)	Fat/TS (g/100 g TS)	NaCl (g/100 g Cheese)
Rennet	Animal ($n = 9$)	4.99 ± 0.03	75.0 ± 3.0	36.1 ± 3.2	48.2 ± 3.3	1.61 ± 0.29
	Vegetable ($n = 9$)	4.97 ± 0.12	72.6 ± 3.3	32.1 ± 3.8	44.2 ± 4.8	1.77 ± 0.16
	F		42.1	144.0	11.83	28.18
	p	ns	0.001	0.001	0.01	0.001
Diet	Control ($n = 6$)	4.90 ± 0.10 [a]	73.6 ± 4.0	35.7 ± 3.2 [a]	48.5 ± 2.4 [a]	1.61 ± 0.04 [a]
	DOP40 ($n = 6$)	5.02 ± 0.03 [b]	73.9 ± 2.9	32.5 ± 5.2 [b]	43.9 ± 5.9 [b]	1.91 ± 0.27 [b]
	DOP80 ($n = 6$)	5.02 ± 0.70 [b]	74.0 ± 3.2	34.2 ± 3.0 [ab]	46.1 ± 3.5 [ab]	2.07 ± 0.24 [c]
	F	36.8		4.82	5.26	26.36
	p	0.001	Ns	0.01	0.05	0.001
Rennet × Diet	F	34.3	128.0	14.2	4.06	10.82
	p	0.001	0.001	0.001	0.05	0.001

Values followed by the same letter within the same column are not significantly different ($p > 0.05$) according to Tukey's multiple range test.

In addition, we found that both rennet type and diet had an influence on the chemical composition of the cheeses. However, other authors [32] found that diet had a greater effect on physicochemical parameters than rennet.

3.2. Odour Sensory Profile

We performed a one-way ANOVA for each sensory attribute with the assessor as the factor. The results of the analyses revealed that the panel worked as a whole (p-value between 0.65 and 0.95 for most attributes).

Table 3 presents the results of the descriptive analysis (mean and standard deviation) and the analysis of variance (rennet × feeding) of odour attributes. The results show that there was a single qualitative profile for the analysed cheeses, with butter, cake, goat, and butyric/propionic acid olfactory notes. Additionally, cheeses from goats fed with a diet based on dried citrus pulp pellets made with vegetable coagulant had toffee and nut olfactory notes. For common sensory attributes, the effect of rennet was significant for butter ($p < 0.05$), cake ($p < 0.05$), goat ($p < 0.001$), and butyric/propionic acid ($p < 0.01$). The effect of diet was significant for overall intensity ($p < 0.001$), butter ($p < 0.01$) and goat ($p < 0.001$), and there was (rennet × feeding) interaction effect for butter ($p < 0.001$), cake ($p < 0.001$), and goat ($p < 0.01$). The cheeses made with animal rennet showed a higher odour intensity for butter, cake, goat, and butyric/propionic acid olfactory notes than vegetable rennet ones. These observations agree with the findings of researchers for Andalusian goat cheeses [14]. Cheeses from goats fed with a diet based on dried citrus pulp pellets had higher overall odour intensity and a greater goat olfactory note than the cheeses from goats fed with the control diet. In addition, these cheeses had toffee and nut olfactory notes.

Table 3. Descriptive measures (mean and standard deviation) and analysis of variance (rennet × feeding) of odour attributes (F-value and probability value).

Odour Attribute	Rennet		Diet		F-Value		*p*-Value
Overall intensity	Animal:	6.2 ± 0.9	Control:	5.7 ± 0.9 [a]	R [1]:		ns
	Vegetable:	6.0 ± 0.8	DOP40:	6.1 ± 0.6 [b]	D [2]:	14.15	0.001
			DOP80:	6.5 ± 0.8 [c]	R*D:		ns
Butter	Animal:	5.3 ± 0.8	Control:	5.1 ± 0.7 [a]	R:	4.18	0.05
	Vegetable:	5.0 ± 1.7	DOP40:	4.8 ± 2.1 [ab]	D:	5.97	0.01
			DOP80:	5.6 ± 0.7 [ac]	R*D:	9.52	0.001
Cake	Animal:	5.0 ± 0.7	Control:	4.8 ± 0.7	R:	5.29	0.05
	Vegetable:	4.7 ± 0.9	DOP40:	4.8 ± 0.9	D:		ns
			DOP80:	5.0 ± 1.0	R*D:	8.07	0.001
Toffee	Animal:	-	Control:	1.2 ± 1.7	R:	-	-
	Vegetable:	1.8 ± 1.3	DOP40:	-	D:	-	-
			DOP80:	1.6 ± 0.7	R*D:	-	-
Nuts	Animal:	-	Control:	-	R:	-	-
	Vegetable:	1.7 ± 0.9	DOP40:	1.3 ± 0.6	D:	-	-
			DOP80:	1.9 ± 1.4	R*D:	-	-
Goat	Animal:	3.0 ± 1.1	Control:	1.7 ± 1.3 [a]	R:	32.40	0.001
	Vegetable:	1.9 ± 1.5	DOP40:	3.6 ± 1.1 [b]	D:	40.52	0.001
			DOP80:	2.1 ± 1.2 [a]	R*D:	5.64	0.01
Butyric/propionic acid	Animal:	5.1 ± 1.0	Control:	5.0 ± 0.8	R:	9.83	0.01
	Vegetable:	4.6 ± 0.8	DOP40:	4.7 ± 0.9	D:		ns
			DOP80:	4.7 ± 1.0	R*D:		ns

[1] R = Rennet; [2] D = Diet. Values followed by the same letter within the same column are not significantly different ($p > 0.05$) according to Tukey's multiple range test.

Unfortunately, very little research has been conducted on the flavour of goat cheeses [14,35,36], and most of these works only give information on the cheeses' basic tastes or trigeminal sensations [28,31,37–39] and not on the odour and aroma attributes characterising them.

3.3. Volatile Compounds

Table 4 lists the volatile compounds (area units, AU; $\times 10^7$) isolated from the cheeses by SPME-GC-MS. We detected 86 compounds: 19 acids, 10 ketones, 14 alcohols, 20 esters, 7 aliphatic hydrocarbons, 6 aromatic compounds, 4 lactones, 2 aldehydes, 3 furanoids, and 1 sulphur compound. Although the concentration of volatile compounds in cheeses made of vegetable rennet was lower, the variability of compounds (particularly esters, aliphatic hydrocarbons, and aldehydes) was higher in the vegetable rennet cheeses. Differences in the enzymatic activity of each type of coagulant may explain this phenomenon. Vegetable coagulant cheeses present slower lipolysis and faster proteolysis rates than those prepared with animal rennet [40,41].

Table 4. Mean and standard deviation of volatile compounds isolated from Payoya goat cheese.

LRI [1]	Volatile Compound	AU [2]	A × C	A × DOP	V × C	V × DOP [3]
Acids		2475.37 ± 1342.30				
1485	Acetic acid	170.54 ± 46.45	×	×	×	×
1568	Propanoic acid	1.99 ± 0.75	×	×	×	×
1594	2-Methyl propanoic acid	8.91 ± 4.13	×	×	×	×
1656	Butanoic acid	921.33 ± 678.06	×	×	×	×
1696	3-Methyl butanoic acid	17.70 ± 20.33	×	×	×	×
1761	Pentanoic acid	6.00 ± 4.10	×	×	×	×
1875	Hexanoic acid	796.30 ± 597.14	×	×	×	×
1978	Heptanoic acid	2.80 ± 1.66	×	×	×	×
2086	Octanoic acid	124.79 ± 64.59	×	×	×	×
2133	4-Methyl octanoic acid	0.58 ± 0.27	×	×	×	×
2177	Nonanoic acid	1.34 ± 0.93	×	×	×	×
2259	Decanoic acid	90.65 ± 146.82	×	×	×	×
2303	Undecanoic acid	2.78 ± 3.47	×	×	×	×
2331	Decenoic acid	4.05 ± 3.70				×
2402	Dodecanoic acid	44.02 ± 13.70	×	×	×	×
2530	Tetradecanoic acid	142.43 ± 271.01		×	×	×
2597	Pentadecanoic acid	36.57 ± 35.49				×
2680	Hexadecanoic acid	170.52 ± 425.29		×	×	×
2714	9-Hexadecenoic acid	35.88 ± 42.40				×
Ketones		223.66 ± 166.11				
867	2-Butanone	3.76 ± 2.51	×	×	×	×
925	4-Hydroxy-2-butanone	13.03 ± 35.47	×	×	×	×
982	2-Pentanone	14.55 ± 10.71	×	×	×	×
1091	2-Methyl-3-pentanone	2.49 ± 1.27	×	×	×	×
1194	2-Heptanone	78.93 ± 63.76	×	×	×	×
1293	2-Octanone	1.74 ± 1.37	×	×	×	×
1301	3-Hydroxy-2-butanone	6.69 ± 3.15	×	×	×	×
1399	2-Nonanone	96.74 ± 91.79	×	×	×	×
1455	8-Nonen-2-one	3.62 ± 3.21	×	×	×	×
1610	2-Undecanone	2.23 ± 1.91	×	×	×	×
Alcohols		210.94 ± 86.75				
936	Ethanol	44.16 ± 19.83	×	×	×	×
1128	Pentan-2-ol	15.83 ± 7.70	×	×	×	×
1137	Methoxyethanol	4.07 ± 2.33	×	×	×	×
1152	Butan-1-ol	3.08 ± 1.30	×	×	×	×
1215	3-Methyl-1-butan-ol	11.42 ± 7.53	×	×	×	×
1326	2- Heptanol	26.90 ± 14.56	×	×	×	×
1362	1-Hexanol	3.40 ± 1.91	×	×	×	×
1427	1-Octen-3-ol	0.57 ± 0.30	×	×	×	×
1527	2-Nonanol	6.64 ± 6.34	×	×	×	×
1556	Propan-1,2-diol	36.42 ± 27.82	×	×	×	×
1576	Hexagol	1.84 ± 1.78	×	×	×	×
1580	Hexa-2,4-dien-1-ol	1.19 ± 0.96	×	×	×	×
1592	Butan-2,3-diol	55.20 ± 36.05	×	×	×	×
1794	5-Ethyl-2-heptanol	1.56 ± 1.58	×	×	×	×
Esters		110.44 ± 43.95				
903	Ethyl acetate	13.17 ± 6.52	×	×	×	×
1044	Ethyl butanoate	36.68 ± 16.01	×	×	×	×
1166	Propyl butanoate	1.78 ± 1.39	×	×	×	×
1227	Butyl butanoate	1.06 ± 0.57	×	×	×	×
1241	Ethyl hexanoate	33.80 ± 21.52	×	×	×	×
1272	3-Methylbutyl 3-methyl butanoate	6.11 ± 7.97	×	×	×	×
1342	Ethyl heptanoate	0.50 ± 0.29	×	×	×	×

Table 4. *Cont.*

LRI [1]	Volatile Compound	AU [2]	A × C	A × DOP	V × C	V × DOP [3]
1356	2-Hydroxy ethyl propanoate	1.21 ± 0.84	×	×	×	×
1382	2-Hydroxy ethyl butanoate	0.66 ± 0.61	×	×	×	×
1421	Butyl hexanoate	0.46 ± 0.22	×	×	×	×
1443	Ethyl octanoate	3.95 ± 1.80	×	×	×	×
1466	Isopentyl hexanoate	1.87 ± 2.19	×	×	×	×
1605	Butyl octanoate	0.47 ± 0.32	×	×	×	×
1648	Ethyl decanoate	2.50 ± 1.00	×	×	×	×
1715	Propyl decanoate	0.73 ± 0.43	×	×	×	×
1853	Ethyl palmitate	1.02 ± 0.60				×
2233	Methyl hexadecanoate	0.96 ± 1.21				×
2430	Methyl octadecenoate	2.30 ± 3.99				×
2440	Ethyl heptadecanoate	5.65 ± 9.79				×
2474	Methyl (Z)-9-octadecenoate	19.00 ± 16.72				×
Aliphatic hydrocarbons		10.28 ± 7.99				
1101	Undecane	3.05 ± 1.32	×	×	×	×
1499	Pentadecane	1.03 ± 0.42	×	×	×	×
1600	Hexadecane	0.94 ± 0.42	×	×	×	×
1800	Octadecane	2.26 ± 2.32	×	×	×	×
1900	Nonadecane	3.88 ± 4.03			×	×
2000	Eicosane	1.76 ± 1.91			×	×
2100	Heneicosane	1.07 ± 0.82				×
Aromatic hydrocarbons		6.09 ± 2.76				
1269	Styrene	0.56 ± 0.33	×	×	×	×
1544	Benzaldehyde	1.13 ± 0.58	×	×	×	×
1666	Phenylacetaldehyde	1.71 ± 1.26	×	×	×	×
1933	2-Phenylethanol	1.53 ± 0.77	×	×	×	×
2035	Phenol	0.64 ± 0.28	×	×	×	×
2108	2-Methylphenol (p-cresol)	0.79 ± 0.46	×	×	×	×
Lactones		4.07 ± 3.00				
1725	δ-Hexalactone	1.67 ± 0.85	×	×	×	×
1941	δ-Octalactone	0.98 ± 0.85	×	×	×	×
1991	δ-Decalactone	1.14 ± 1.68	×	×	×	×
2369	δ-Dodecalactone	1.72 ± 1.96				×
Aldehydes		3.27 ± 4.19				
2040	Pentadecanal	1.19 ± 0.62				×
2140	Hexadecanal	2.67 ± 3.55				×
Furans		2.57 ± 1.10				
1636	5-Methyl-2-furfural	0.75 ± 0.25	×	×	×	×
1679	2-Furanmethanol	0.97 ± 0.50	×	×	×	×
2055	Dihydro-5-phenyl-2(3H)-furanone	0.95 ± 0.64	×	×	×	×
Sulphur compounds		1.38 ± 0.50				
1932	Dimethylsulphone	1.38 ± 0.50	×	×	×	×

[1] Linear retention index; [2] Area units (×10⁷); [3] A: Animal rennet; C: Control diet; DOP: Dry orange pulp diet including both DOP40 and DOP80; V: Vegetable rennet. Compounds marked with "×" were found in samples of each rennet × diet combination.

Short-chain fatty acids were the most abundant volatile compounds of all the identified fractions. Butanoic, hexanoic, and acetic acids had the highest percentages in the volatile fraction of Payoyo cheese (the cheese from Payoya goats milk), in decreasing order. Acids also play a predominant role in the aroma of many other goat cheeses, such as Ibores, Majorero, Palmero, Sepet, Xinotyri, and Sainte-Maure [42–47]. Free fatty acids containing two or more carbon atoms may originate from lipolysis, proteolysis, or the degradation of lactose. The source of these enzymatic activities can be starter cultures, moulds, or indigenous milk enzymes [48]. The amount of total acids increased during ripening owing to the fat hydrolysis process. However, shorter fatty acids can also be produced by the oxidation of ketones, esters, and aldehydes. The absence of starter cultures during manufacturing, together with the low amount of aldehydes present in the cheeses (qualitatively and quantitatively), suggests that acetic or propionic acids may have been derived from the oxidation of the corresponding aldehydes.

Methylketones were the most abundant type within this fraction of volatile compounds, as in other goat varieties and surface-ripened cheeses [42,48]. Methylketones are precursors of secondary alcohols in the ß-oxidation of free fatty acids [49], and they have low perception thresholds. Two ketones were especially abundant: 2-nonanone and 2-heptanone. Similar results were also found in

Spanish PDO raw milk cheeses [50,51], and so they may play a key role in the final aroma of raw milk cheeses in general.

Butan-2,3-diol and ethanol were the main alcohols detected in Payoyo cheese. Butan-2,3-diol is the intermediate product of the reduction of diacetyl to acetoin by bacterial enzymes present in raw milk; the compound can, in turn, be reduced to butan-2-one and finally to butan-2-ol [46]. Acetoin and butan-2-one were present in our cheese and in other raw milk cheeses, while the diacetyl itself and its final degradation product, butan-2-ol, were absent. Ethanol is derived from the fermentation of lactose and from the catabolism of amino acids such as alanine and plays a fundamental role in the formation of esters [49,52]. This alcohol is predominant in a large number of goat cheeses [46,53].

Although less abundant in our samples, the ester fraction presented the highest variability in Payoya goat cheeses, encompassing 20 compounds. Esters result from the reaction between fatty acids (short- and medium-chain) and secondary alcohols that come from lactose degradation or from amino acid catabolism [48]. Ethyl esters represented half of the compounds detected within this chemical family. They play a remarkable role in the aroma profile of cheese due to their low perception thresholds.

Several minority compounds making up 1% of the total volatile fraction were identified in Payoyo cheeses, including hydrocarbons, lactones, aldehydes, furans, and sulphur compounds. Although seven alkanes were identified in the Payoyo cheeses, their high odour thresholds make them insignificant contributors to cheese aroma. However, these compounds are crucial to form other aromatic compounds by acting as precursors in various degradation pathways [54]. Concerning aromatic hydrocarbons, phenylacetaldehyde, benzaldehyde, and 2-phenylethanol were identified at higher concentrations. McSweeney and Sousa [49] suggested that phenylacetaldehyde can be formed by the Strecker reaction from phenylalanine and acetaldehyde derived from threonine. Afterwards, benzaldehyde may be produced from the α-oxidation of phenylacetaldehyde or from ß-oxidation of cinnamic acid. Both phenylacetaldehyde and benzaldehyde have also been detected in other cheeses made with raw milk, such as Xinotyri and Torta de la Serena cheese [46,50]. Four δ-lactones were identified in our samples. In cheese, lactones are the result of a lactonization after the hydrolysis of hydroxy-fatty acid triglycerides. As a result, the concentration of lactones usually correlates with the extent of lipolysis, which is consistent with our results. Aldehydes derive from the conversion of amines and α-ketoacids originating from the catabolism of amino acids and are rapidly reduced to alcohols or oxidised to acids. Therefore, they do not accumulate to high concentrations, and their presence is not significant in the volatile profile of cheese [49]. Only two long straight-chain aldehydes were detected in Payoya goat cheese: hexadecanal and pentadecanal. These compounds have relatively high perception thresholds and are probably unimportant. Three furan compounds, including 2-furanmethanol, were detected in Payoyo cheeses and in a variety of goat cheeses including Flor de Guía [55] and Xynotyri [46]. However, this furan fraction, together with one sulphur compound, were detected in very low concentrations in our cheeses.

3.4. Variability and Correlation of Payoya Goat Cheese Volatile Compounds and Odour Attributes

Odour and flavour descriptors associated with volatile compounds detected in Payoya goat cheeses are presented in Table 5. The conversion] of triglycerides to fatty acids and glycerol by enzymatic hydrolysis (lipolysis) is essential to flavour development in many cheese varieties [49]. Fatty acids are not only key aroma contributors themselves but are also precursors of many other crucial compounds [48]. Short-chain fatty acids like acetic and propanoic acids typically have vinegar, sour, or pungent odours [45,56,57]. Straight medium-chain fatty acids contribute significantly to the aroma of many cheese types [48], producing slight rancid cheese-like notes. However, high concentrations of these fatty acids can produce undesirable attributes. Other members of this chemical family, such as odd-numbered-chain fatty acids (heptanoic and nonanoic acids), impart a goat flavour to goat cheese [45,48,56]. This potent odour is also caused by branched-chain fatty acids present in Payoya goat cheese, like 4-methyl octanoic acid [45]. Although its concentration was moderate, 3-methyl butanoic acid was also present in our samples. This fatty acid is derived from leucine amino acid breakdown and is related to very-ripe-cheese aroma due to the rancid, cheesy, sweaty, and putrid odours it imparts [57,58].

Table 5. Volatile compounds associated with sensory descriptors.

Volatile Compound	Sensory Descriptor	References [1]
Acetic acid	Vinegar, sour, pungent, peppers, green, floral	[45,56,57]
Propanoic acid	Pungent, sour milk, cheese, gas, burnt, cloves, fruity	[57]
2-Methylpropanoic acid	Nutty, cheesy, rancid, butter	[48,59]
Butanoic acid	Cheesy, sharp, rancid, rennet, brine	[45,56,59]
3-Methylbutanoic acid	Acidic, cheese, sweaty, rancid, unpleasant	[57,58]
Pentanoic acid	Rancid yeast, unpleasant fermented	[60]
Hexanoic acid	Goaty, sweaty, rancid, cheesy, sharp	[45,56,57]
Heptanoic acid	Goaty, cheesy, sweaty, rancid	[45,56]
Octanoic acid	Waxy, sweaty, soapy, cheesy, rancid, pungent	[45,56,57]
4-Methyl octanoic acid	Goaty, sour	[45]
Nonanoic acid	Goaty	[48]
Decanoic acid	Sour, waxy, fatty, soapy	[45,56]
Dodecanoic acid	Soapy	[45]
Tetradecanoic acid	Sweaty, animal	[48]
Hexadecanoic acid	Waxy, lard, tallow	[57]
Butan-2-one	Milky, toasty, sweet, ether-like, slightly nauseating notes	[57,59,61]
2-Pentanone	Sweet, fruity, orange peel, caramel, butter, creamy	[56,58,61]
2-Methyl-3-pentanone	Candy	[48]
2-Heptanone	Musty, soapy, blue cheese	[56,58]
2-Octanone	Fruity	[48]
3-Hydroxy 2-butanone	Buttery, sour milk, milky, toasty	[45,57,59]
2-Nonanone	Fatty, floral, musty, fruity, soapy, malty, rotten fruit, hot milk, green, earthy notes	[45,56,61]
Non-8-en-2-one	Blue cheese	[48]
Ethanol	Alcohol notes, dry dust	[45,61]
Pentan-2-ol	Alcohol, fruity, green, fresh	[56,57]
Butan-1-ol	Banana-like, wine-like, fusel oil	[57]
3-Methyl-1-butan-ol	Fresh cheese, breathtaking, alcoholic, fruity, grainy, solvent-like	[57,61]
2-Heptanol	Fruity, sweet, green, earthy, dry, dusty carpet	[56,57]
1-Hexanol	Flowery, fruity	[56]
1-Octen-3-ol	Mushroom-like, mouldy, earthy	[57]
2-Nonanol	Fatty green	[57]
Ethyl acetate	Solvent, fruity, pineapple	[57]
Ethyl butanoate	Fruity, apple-like, sweet, chewing gum, green, banana	[45,56–58]
Propyl butanoate	Fruity, sweet, pineapple-like	[56]
Butyl butanoate	Nutty	[59]
Ethyl hexanoate	Orange, sour, fruity, apple-like, mouldy, rennet, brine, sweet, green fermented	[45,56,58,59]
Butyl hexanoate	Fruity, pineapple-like, mouldy	[56]
Ethyl octanoate	Fruity, winey, pear, apricot, sweet, banana, pineapple	[45,56,57]
Ethyl decanoate	Fruity, winey, fatty	[45,56]
Benzaldehyde	Bitter almond, sweet cherry	[61]
Phenylacetaldehyde	Flower, hyacinth, honey-like, rosey, violet-like, styrene	[57,58]
2-Phenylethanol	Sweet-flowery, rose	[60]
2-Methylphenol (p-cresol)	Phenolic, medicinal, cowy, barny, musty, stable	[48]
δ-Octalactone	Coconut-like, fruity, peach-like	[48]
δ-Decalactone	Peach, coconut-like, creamy, milk fat	[45,57]
δ-Dodecalactone	Coconut, cheesy, sweet, soapy, buttery, peach, milk fat	[45,57]
Dimethylsulphone	Sulphurous, hot milk, burnt	[48]

[1] Literature references where volatile compounds were previously identified in cheese.

Ketones have low perception thresholds and contribute to the pungent aroma of blue cheeses. However, butan-2-one, which has a milky, toasty, and sweet odour, was identified as a main odorant of cheddar cheese in moderate concentrations. Moreover, 2-heptanone and 2-nonanone, with musty and soapy odours, both of them, are important compounds in creamy and natural Emmental and

Gorgonzola cheeses [48]. Fruity, floral, and musty notes are associated with other lactones, including octan-2-one and nonan-2-one.

The presence of branched-chain primary alcohols such as 3-methyl-butan-1-ol indicates the reduction of the corresponding aldehyde from the isoleucine amino acid. 3-Methyl-butan-1-ol has also been identified in other goat cheeses [43,47] and imparts pleasant notes to fresh cheese [57,61]. However, Garde et al. [62] considered this compound undesirable due to its association with barnyard and animal flavours.

Ethyl esters provide floral and fruity notes to cheese odours when they are present in low concentrations but yeasty notes when present in high concentrations [45,56,58,59]. The increase of esters may be associated with the decline of some alcohols at the end of ripening as the result of bacterial and yeast activity. However, no starter cultures were used during Payoyo cheese manufacturing, and so the esterification of alcohols by these agents was limited along with the undesirable odours. In addition, methyl esters may contribute to the cheese aroma by minimising the sharp aroma of fatty acids [46].

Regarding the minority compounds detected in Payoyo cheese, aromatic hydrocarbons such as benzaldehyde and phenylacetaldehyde provide sweet, floral, and fruity notes, while 2-methylphenol is associated with cowy, barny, musty, and stable odours [48,57,58,61]. Lactones are characteristic coconut-like odorants in cheeses, and dimethylsulphone adds sulphurous, hot milk, and burnt odours [45,48,57].

PCA was performed on sensory attributes and volatile compounds. Payoyo cheese was well-differentiated by rennet, but not by diet (Figure 1). Volatile compounds were selected for PCA depending on their chemical nature and possible impact as odour-active compounds in Payoya goat cheese. Six principal components accounting for 86.6% of the total variance defined the variation in the odour among different cheeses. Cheeses made from vegetable coagulant tended to receive high negative scores on the PC1 axis, which explained 35.83% of the variance. Within the vegetable rennet group, the C and DOP40 cheeses were separated from DOP80 diet cheeses by PC2. These distinctions were not evident in cheeses made of animal rennet, which had a much more homogeneous distribution.

The variables with high loadings (higher than 0.5) on PC1 included goat and toffee odours, all the acids included in the analysis, and the aromatic compounds. Regarding PC2 (which explained 19.5% of the variance), the variables with high loadings were again goat and toffee odours, one lactone, and one ester. Cheeses made of animal rennet appeared associated with acids (linear and branched) and aromatic compounds such as 2-methylphenol or benzaldehyde. All the samples made with milk from goats fed with the control diet, as well as one sample from DOP40 goats, and one from DOP80 goats, were closer to goat and pungent odours; the remaining samples were closer to cake and butter odours. On the other hand, vegetable rennet-made cheeses were linked to toffee odour, esters, and δ-dodecalactone compounds (particularly those made with the milk of goats fed with the DOP80 diet). Some compounds, such as octanoic acid, 2-methyl propanoic acid, and benzaldehyde, were not represented by the PCA as close to the odours they typically provide (goaty, nutty and bitter almond/sweet cherry, respectively). Thus, the separation of volatile compounds in cheeses did not follow exactly the same pattern as the separation of sensory analysis. This result agreed with authors such as Hannon et al. [63] who have suggested that volatile compounds detected following purge-and-trap extraction contribute only partially to the perception of flavour in the final cheese. In addition, the relationship between chemical compounds and perceived aromas and flavours is still unclear due to the lack of direct linear relationships between compounds and perceptions [64]. This means that statistically associated variables do not imply a causative relationship.

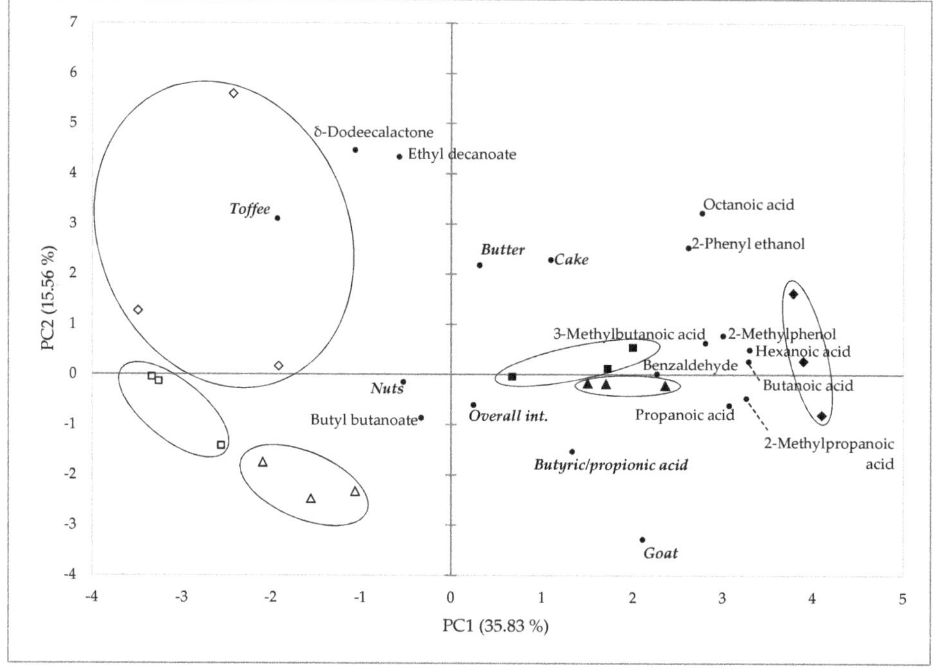

Figure 1. Principal component analysis plot representing the differentiation of Payoyo cheeses made with the milk of goats fed with different diets and rennet, based on the main volatile compounds and sensory attributes. ▲, control diet × animal rennet; △, control diet × vegetable rennet; ■, DOP40 diet × animal rennet; □, DOP40 diet × vegetable rennet; ◆, DOP80 diet × animal rennet; ◇, DOP80 diet × vegetable rennet.

4. Conclusions

The artisanal products analysed in this experiment retained the characteristics of goat cheeses. The use of citrus by-product in Payoya goat feeding did not substantially affect the physicochemical analysis, olfactory attributes, or volatile profiles of the cheeses. Dried citrus pulp can be used as a substitute for cereal in concentrates without affecting the distinctive final characteristics of these ripened raw goat milk cheeses. In addition, this dietary strategy may increase the value of a by-product of the agri-food orange juice industry.

The association between odour descriptors and volatile composition was not as clear as expected. Odour characteristics in complex matrices such as cheeses depend not only on combinations of volatiles but also on interactions between specific compositional variables. The relationship between sensory and volatile profiles was not entirely conclusive in this study.

Author Contributions: Conceptualisation: M.D.P., J.L.G., and C.A.R.; methodology: M.D.P., H.G.S., P.R.P.-C., O.P.P., J.L.G., L.Á.Z., and C.A.R.; formal analysis: H.G.S., P.R.P.-C., O.P.P., and C.A.R.; data curation: H.G.S., P.R.P.-C., O.P.P.; and C.A.R.; writing—original draft preparation: H.G.S., P.R.P.-C., O.P.P., and C.A.R.; writing—review and editing: H.G.S., P.R.P.-C., O.P.P., C.A.R., M.D.P., J.L.G., and L.Á.Z.; supervision, project administration and funding acquisition: J.L.G., M.D.P., and L.Á.Z. All authors have read and agreed to the published version of the manuscript.

Funding: This research was supported by a Regional Project funded by Excma. Diputación de Huelva, Spain.

Acknowledgments: The authors are grateful to the Excma. Diputación Provincial de Huelva (Spain) for their financial support, Cítricos del Andévalo, S.A. (García Carrión) for supplying pellets of dehydrated orange pulp, and OVIPOR, Soc.Coop. The authors wish to thank the farm staff of Huelva University for their technical support and Quesos Doñana, S.L for cheese manufacture.

Conflicts of Interest: The authors declare no conflict of interest. The funding sponsors had no role in the design of the study; the collection, analysis or interpretation of data; the writing of the manuscript; or the decision to publish the results.

References

1. Castel, J.M.; Mena, Y.; Ruiz, F.A.; Camúñez-Ruiz, J.; Sanchez-Rodriguez, M. Changes occurring in dairy goat production systems in less favoured areas of Spain. *Small Rumin. Res.* **2011**, *96*, 83–92. [CrossRef]

2. Gutierrez-Peña, R.; Fernández-Cabanás, V.M.; Mena, Y.; Delgado-Pertíñez, M. Fatty acid profile and vitamins A and E contents of milk in goat farms under Mediterranean wood pastures as affected by grazing conditions and seasons. *J. Food Compos. Anal.* **2018**, *72*, 122–131. [CrossRef]

3. Menéndez-Buxadera, A.; Molina, A.; Arrebola, F.; Clemente, I.; Serradilla, J.M. Genetic variation of adaptation to heat stress in two Spanish dairy goat breeds. *J. Anim. Breed. Genet.* **2012**, *129*, 306–315. [CrossRef] [PubMed]

4. López, M.C.; Estellés, F.; Moya, V.J.; Fernández, C. Use of dry citrus pulp or soybean hulls as a replacement for corn grain in energy and nitrogen partitioning, methane emissions, and milk performance in lactating Murciano-Granadina goats. *J. Dairy Sci.* **2014**, *97*, 7821–7832. [CrossRef]

5. Guzmán, J.L.; Delgado-Pertíñez, M.; Beriain, M.J.; Pino, R.; Zarazaga, L.Á.; Horcada, A. The use of concentrates rich in orange by-products in goat feed and its effects on physico-chemical, textural, fatty acids, volatile compounds and sensory characteristics of the meat of suckling kids. *Animals* **2020**, *10*, 766. [CrossRef]

6. Ares, J.L.; de Asis Ruiz Morales, F.; Barriga, D.; Berrocal, J.; Calvente, I.; Cárdenas, J.M.; Carrasco, C.; Castel, J.M.; Galán, H.; Gámez, M.; et al. *Patrimonio Quesero Andaluz: Quesos de Andalucía*; Grupo de Desarolo Rural Valle del Guadalhorce: Málaga, Spain, 2009; ISBN 978-84-613-6388-9.

7. Shah, M.A.; Mir, S.A.; Paray, M.A. Plant proteases as milk-clotting enzymes in cheesemaking: A review. *Dairy Sci. Technol.* **2014**, *94*, 5–16. [CrossRef]

8. Pulina, G.; Milán, M.J.; Lavín, M.P.; Theodoridis, A.; Morin, E.; Capote, J.; Thomas, D.L.; Francesconi, A.H.D.; Caja, G. Invited review: Current production trends, farm structures, and economics of the dairy sheep and goat sectors. *J. Dairy Sci.* **2018**, *101*, 6715–6729. [CrossRef]

9. FAO, 2018 FAOSTAT. Available online: http://www.fao.org/faostat/es/#data/QC (accessed on 28 July 2020).

10. Salvador, A.; Igual, M.; Contreras, C.; Martínez-Navarrete, N.; del Mar Camacho, M. Effect of the inclusion of citrus pulp in the diet of goats on cheeses characteristics. *Small Rumin. Res.* **2014**, *121*, 361–367. [CrossRef]

11. de la Haba Ruiz, M.A.; Ruiz Pérez-Cacho, P.; Dios-Palomares, R.; Galán-Soldevilla, H. Classification of artisanal Andalusian cheeses on physicochemical parameters applying multivariate statistical techniques. *Dairy Sci. Technol.* **2016**, *96*, 95–106. [CrossRef]

12. ISO/IDF ISO/IDF International Organization for Standardization/International Dairy Federation (2008). In *Cheese—Determination of Fat Content—Van Gulik Method*; ISO 3433:2008 (IDF 222:2008); ISO: Genève, Switzerland, 2008.

13. AOAC International (Ed.) *Official Methods of Analysis of AOAC International*, 16th ed.; AOAC International: Arlington, VA, USA, 1999; Volume 1, ISBN 0-935584-54-4.

14. Ruiz Pérez-Cacho, P.; Ruiz, M.; Dios-Palomares, R.; Galán-Soldevilla, H. Linear regression models for estimating the effect of technological factors on the sensory characteristics of goat cheeses. *Int. J. Food Sci. Technol.* **2019**, *54*, 2396–2407. [CrossRef]

15. International Organization for Standardization. *International Standard 5492. Sensory Analysis. Sensory Vocabulary*; ISO 5492:2008 ISO.; ISO: Genève, Switzerland, 2008.

16. International Organization for Standardization. *International Standard 8586. Sensory Analysis. Methodolog. General Guidance for the Selection, Training and Monitoring of Selected Assessors and Expert Sensory Assessors*; ISO 8586:2012 ISO.; ISO: Genève, Switzerland, 2012.

17. International Organization for Standardization. *International Standard 13299. Sensory Analysis. Methodology. General Guidance for Establishing a Sensory Profile*; ISO 13299:2016 ISO.; ISO: Genève, Switzerland, 2016.

18. Galán-Soldevilla, H.; Ruiz Pérez-Cacho, M.P.; Serrano Jiménez, S.; Jodral Villarejo, M.; Manzanares, A.B. Development of a preliminary sensory lexicon for floral honey. *Food Qual. Prefer.* **2005**, *16*, 71–77. [CrossRef]

19. Ruiz Pérez-Cacho, M.P.; Galán-Soldevilla, H.; León Crespo, F.; MolinaRecio, G. Determination of the sensory attributes of a Spanish dry-cured sausage. *Meat Sci.* **2005**, *71*, 620–633. [CrossRef] [PubMed]

20. Ruiz Pérez-Cacho, P.; Galán-Soldevilla, H.; Mahattanatawee, K.; Elston, A.; Rouseff, R. Sensory lexicon for fresh squeezed and processed orange juices. *Food Sci. Technol. Int.* **2008**, *14*, 131–141. [CrossRef]
21. Rodríguez, M.I.; Serrano, S.; Galán-Soldevilla, H.; Ubera, J.; Jodral, M. Characterisation of Sierra Morena citrus blossom honey (Citrus sp). *Int. J. Food Sci. Technol.* **2010**, *45*, 2008–2015. [CrossRef]
22. Galán-Soldevilla, H.; Pérez-Cacho, P.; Campuzano, J. Determination of the characteristic sensory profiles of Aloreña table-olive. *Grasas Aceites* **2013**, *64*, 442–452. [CrossRef]
23. Rodríguez, M.I.; Serrano, S.; Galán-Soldevilla, H.; Piva, G.; Ubera, J. Sensory analysis integrated by palynological and physicochemical determinations plays a key role in differentiating unifloral honeys of similar botanical origins (*Myrtaceaehoneys* from southern Spain). *Int. J. Food Sci. Technol.* **2015**, *50*, 1545–1551. [CrossRef]
24. Araujo, D.; Pérez-Cacho, P.; Serrano, S.; Dios-Palomares, R.; Galán-Soldevilla, H. Sensory profile and physico-chemical properties of artisanal honey from Zulia, Venezuela. *Foods* **2020**, *9*, 339. [CrossRef]
25. International Organization for Standardization. *International Standard 8589. Sensory Analysis. General Guidance for the Design of Test Rooms*; ISO 8589:2007 ISO.; ISO: Genève, Switzerland, 2007.
26. International Organization for Standardization. *International Standard 22935-1. Milk and Milk Products. Sensory Analysis. Part 1: General Guidance for the Recruitment, Selection, Training and Monitoring of Assessors*; ISO 22935-1:2009 ISO.; ISO: Genève, Switzerland, 2009.
27. International Organization for Standardization. *International Standard 22935-2. Milk and Milk Products. Sensory Analysis. Part 2: General Guidance for the Recruitment, Selection, Training and Monitoring of Assessors*; ISO 22935-2:2009 ISO.; ISO: Genève, Switzerland, 2009.
28. Álvarez, S.; Fresno, M.; Méndez, P.; Castro, N.; Fernández, J.R.; Sampelayo, M.R.S. Alternatives for improving physical, chemical, and sensory characteristics of goat cheeses: The use of arid-land forages in the diet. *J. Dairy Sci.* **2007**, *90*, 2181–2188. [CrossRef]
29. Mas, M.; Tabla, R.; Moriche, J.; Roa, I.; Gonzalez, J.; Rebollo Feria, J.; Cáceres, P. Ibores goat's milk cheese: Microbiological and physicochemical changes throughout ripening. *Lait* **2002**, *82*, 579–587. [CrossRef]
30. Tejada, L.; Abellán, A.; Prados, F.; Cayuela, J. Compositional characteristics of Murcia al Vino goat's cheese made with calf rennet and plant coagulant. *Int. J. Dairy Technol.* **2008**, *61*, 119–125. [CrossRef]
31. Fresno, M.; Álvarez, S. Chemical, textural and sensorial changes during the ripening of Majorero goat cheese. *Int. J. Dairy Technol.* **2012**, *65*, 393–400. [CrossRef]
32. Fresno Baquero, M.; Álvarez Ríos, S.; Rodríguez Rodríguez, E.M.; Díaz Romero, C.; Darias Martín, J. Influence of diet and rennet on the composition of goats' milk and cheese. *J. Dairy Res.* **2011**, *78*, 250–256. [CrossRef] [PubMed]
33. González-Fandos, E.; Sanz, S.; Olarte, C. Microbiological, physicochemical and sensory characteristics of Cameros cheese packaged under modified atmospheres. *Food Microbiol.* **2000**, *17*, 407–414. [CrossRef]
34. Peláez Puerto, P.; Fresno Baquero, M.; Rodríguez Rodríguez, E.M.; Darías Martín, J.; Díaz Romero, C. Chemometric studies of fresh and semi-hard goats' cheeses produced in Tenerife (Canary Islands). *Food Chem.* **2004**, *88*, 361–366. [CrossRef]
35. Retiveau, A.; Chambers, D.H.; Esteve, E. Developing a lexicon for the flavor description of French cheeses. *Food Qual. Prefer.* **2005**, *16*, 517–527. [CrossRef]
36. Chambers, D.; Esteve, E.; Retiveau Krogmann, A. Effect of milk pasteurization on flavor properties of seven commercially available french cheese types. *J. Sens. Stud.* **2010**, *25*, 494–511. [CrossRef]
37. Cabezas, L.; Poveda, J.M.; Sánchez, I.; Palop, M.L. Physico-chemical and sensory characteristics of Spanish goat cheeses. *Milchwissenschaft* **2005**, *60*, 48–51.
38. García, V.; Rovira Garbavo, S.; Boutoial, K.; Ferrandini, E.; López, M. Physicochemical, microbiological, textural and sensory changes during the ripening of pasteurised goat milk cheese made with plant coagulant (*Cynara scolymus*). *Int. J. Dairy Technol.* **2015**, *69*, 96–102. [CrossRef]
39. Tejada, L.; Abellán, A.; Cayuela, J.; Martínez-Cacha, A. Sensorial characteristics during ripening of the murcia al vino goat's milk cheese: The effect of the type of coagulant used and the size of the cheese. *J. Sens. Stud.* **2006**, *21*, 333–347. [CrossRef]
40. Rincón Rincón, A.; Pino, V.; Fresno, M.; Jiménez, A.; Álvarez, S.; Ayala, J.; Afonso, A. Influence of vegetable coagulant and ripening time on the lipolytic and sensory profile of cheeses made with raw goat milk from Canary breeds. *Food Sci. Technol. Int.* **2017**, *23*, 254–264. [CrossRef]

41. Ordiales, E.; Benito, M.J.; Martín, A.; Fernández, M.; Hernández, A.; de Guia Córdoba, M. Proteolytic effect of *Cynara cardunculus* rennet for use in the elaboration of 'Torta del Casar' cheese. *J. Dairy Res.* **2013**, *80*, 429–438. [CrossRef] [PubMed]

42. Delgado, F.J.; González-Crespo, J.; Cava, R.; Ramírez, R. Formation of the aroma of a raw goat milk cheese during maturation analysed by SPME–GC–MS. *Food Chem.* **2011**, *129*, 1156–1163. [CrossRef] [PubMed]

43. Castillo, I.; Calvo, M.V.; Alonso, L.; Juárez, M.; Fontecha, J. Changes in lipolysis and volatile fraction of a goat cheese manufactured employing a hygienized rennet paste and a defined strain starter. *Food Chem.* **2007**, *100*, 590–598. [CrossRef]

44. Guillén, M.D.; Ibargoitia, M.L.; Sopelana, P.; Palencia, G.; Fresno, M. Components detected by means of solid-phase microextraction and gas chromatography/mass spectrometry in the headspace of artisan fresh goat cheese smoked by traditional methods. *J. Dairy Sci.* **2004**, *87*, 284–299. [CrossRef]

45. Ercan, D.; Korel, F.; Karagül Yüceer, Y.; Kınık, Ö. Physicochemical, textural, volatile, and sensory profiles of traditional Sepet cheese. *J. Dairy Sci.* **2011**, *94*, 4300–4312. [CrossRef]

46. Bontinis, T.G.; Mallatou, H.; Pappa, E.; Massouras, T.; Alichanidis, E. Study of proteolysis, lipolysis and volatile profile of a traditional Greek goat cheese (Xinotyri) during ripening. *Small Rumin. Res.* **2012**, *105*, 193–201. [CrossRef]

47. Le Quere, J.L.; Pierre, A.; Riaublanc, A.; Demaizières, D. Characterization of aroma compounds in the volatile fraction of soft goat cheese during ripening. *Lait* **1998**, *78*, 279–290. [CrossRef]

48. Curioni, P.M.G.; Bosset, J.O. Key odorants in various cheese types as determined by gas chromatography-olfactometry. *Int. Dairy J.* **2002**, *12*, 959–984. [CrossRef]

49. McSweeney, P.; Sousa-Gallagher, M.J. Biochemical pathways for the production of flavour compounds in cheeses during ripening: A review. *Lait* **2000**, *80*, 293–324. [CrossRef]

50. Carbonell, M.; Nuñez, M.; Fernández-García, E. Evolution of the volatile components of ewe raw milk La Serena cheese during ripening. Correlation with flavour characteristics. *Lait* **2002**, *82*, 683–698. [CrossRef]

51. Delgado, F.J.; González-Crespo, J.; Cava, R.; García-Parra, J.; Ramírez, R. Characterisation by SPME–GC–MS of the volatile profile of a Spanish soft cheese P.D.O. Torta del Casar during ripening. *Food Chem.* **2010**, *118*, 182–189. [CrossRef]

52. Molimard, P.; Spinnler, H.E. Review: Compounds involved in the flavor of surface mold-ripened cheeses: Origins and properties. *J. Dairy Sci.* **1996**, *79*, 169–184. [CrossRef]

53. Massouras, T.; Pappa, E.; Mallatou, H. Headspace analysis of volatile flavour compounds of Teleme cheese made from sheep and goat milk. *Int. J. Dairy Technol.* **2006**, *59*, 250–256. [CrossRef]

54. Gatzias, I.S.; Karabagias, I.K.; Kontominas, M.G.; Badeka, A.V. Geographical differentiation of feta cheese from northern Greece based on physicochemical parameters, volatile compounds and fatty acids. *LWT* **2020**, *131*, 109615. [CrossRef]

55. Gonzalez-Mendoza, L.A.; Diaz-Rodriguez, F. Sensory analysis using sniffing, for determination of the aroma of Flor de Guia cheese. *Aliment. Equipos Tecnol.* **1993**, *12*, 45–49.

56. Abilleira, E.; Schlichtherle-Cerny, H.; Virto, M.; de Renobales, M.; Barron, L.J.R. Volatile composition and aroma-active compounds of farmhouse Idiazabal cheese made in winter and spring. *Int. Dairy J.* **2010**, *20*, 537–544. [CrossRef]

57. Kilcawley, K.N. Cheese Flavour. In *Fundamentals of Cheese Science*; Springer: Boston, MA, USA, 2017; ISBN 978-1-4899-7681-9.

58. Varming, C.; Andersen, L.T.; Petersen, M.A.; Ardö, Y. Flavour compounds and sensory characteristics of cheese powders made from matured cheeses. *Int. Dairy J.* **2013**, *30*, 19–28. [CrossRef]

59. Barron, L.J.R.; Redondo, Y.; Aramburu, M.; Gil, P.; Pérez-Elortondo, F.J.; Albisu, M.; Nájera, A.I.; de Renobales, M.; Fernández-García, E. Volatile composition and sensory properties of industrially produced Idiazabal cheese. *Int. Dairy J.* **2007**, *17*, 1401–1414. [CrossRef]

60. Sádecká, J.; Kolek, E.; Pangallo, D.; Valík, L.; Kuchta, T. Principal volatile odorants and dynamics of their formation during the production of May Bryndza cheese. *Food Chem.* **2014**, *150*, 301–306. [CrossRef]

61. Bertuzzi, A.; McSweeney, P.; Rea, M.; Kilcawley, K. Detection of volatile compounds of cheese and their contribution to the flavor profile of surface-ripened cheese. *Compr. Rev. Food Sci. Food Saf.* **2018**, *17*, 371–390. [CrossRef]

62. Garde, S.; Ávila, M.; Medina, M.; Nuñez, M. Influence of a bacteriocin-producing lactic culture on the volatile compounds, odour and aroma of Hispánico cheese. *Int. Dairy J.* **2005**, *15*, 1034–1043. [CrossRef]

63. Hannon, J.A.; Kilcawley, K.N.; Wilkinson, M.G.; Delahunty, C.M.; Beresford, T.P. Production of ingredient-type cheddar cheese with accelerated Flavor development by addition of enzyme-modified cheese powder. *J. Dairy Sci.* **2006**, *89*, 3749–3762. [CrossRef]
64. Chambers, E.I.V.; Koppel, K. Associations of volatile compounds with sensory aroma and flavor: The complex nature of flavor. *Molecules* **2013**, *18*, 4887–4905. [CrossRef] [PubMed]

Article

Effect of Two Organic Production Strategies and Ageing Time on Textural Characteristics of Beef from the Retinta Breed

Susana García-Torres [1,*], Adoración López-Gajardo [1], David Tejerina [1], Estrella Prior [1], María Cabeza de Vaca [1] and Alberto Horcada [2]

[1] Meat Quality Area, Centro de Investigaciones Científicas y Tecnológicas of Extremadura (CICYTEX-La Orden), Junta de Extremadura, Guadajira, 06187 Badajoz, Spain; dologa19@hotmail.com (A.L.-G.); tejerinabarrado@yahoo.es (D.T.); yeyiprior@hotmail.com (E.P.); merycv@hotmail.com (M.C.d.V.)

[2] Departamento de Ciencias Agroforestales, Escuela Técnica Superior de Ingeniería Agronómica, Universidad de Sevilla, 41013 Sevilla, Spain; albertohi@us.es

* Correspondence: garsus15@hotmail.com

Received: 9 August 2020; Accepted: 27 September 2020; Published: 7 October 2020

Abstract: The primary aim of this paper is to determine the influence of two organic production systems, organic grass-fed (OG) and organic concentrate-fed (OC), vs. a conventional concentrate-fed (CC) system; the second aim is to determine the influence of the ageing period on the physical parameters and texture properties of beef from the Retinta breed. Muscle samples from *Longissimus thoracis* were stored at 2–4 ± 1 °C for 0, 7, 14, and 21 days for the purposes of ageing. Analyses of pH, water losses (drip loss and cooking loss), Warner-Bratzler shear force, texture profile analysis (TPA), and histological analysis of muscle fibre were carried out. The results revealed that organic meat experienced lower drip loss and higher cooking loss than conventional meat. Although the meat of organic grass-fed animals was tougher initially, it showed a higher tenderisation speed in the first ageing days than OC and CC meats. The sarcomere length increased during the ageing period, which showed a negative correlation to shear force. According to its texture characteristics, the Retinta meat produced in organic systems could be recommended by its quality.

Keywords: organic beef; ageing; tenderisation speed; meat quality; sarcomere

---

1. Introduction

In recent years, organic production systems have gained increasing economic importance in the meat sector. This is due to the increasing commercial and social acceptance of these kinds of products, which is mainly associated to aspects relating to food safety, food quality, animal welfare, and environmental benefits [1].

Organic food consumers are willing to pay a premium for organic products, as they see themselves contributing to environmental sustainability and higher standards of animal welfare [2]. Furthermore, organic products are perceived as healthier, which is an item of added value, increasing consumers' willingness to pay a premium for them [3]. Napolitano et al. [4] reported that consumers are influenced by the available information on organic production and that they value not only meat quality, but also the ethical considerations associated with organic farming. Furthermore, Lee and Yun [2] suggested that hedonic attitudes, perceptions of nutritional content, organic welfare, and sensory appeal attributes lead to a positive intention to purchase organic meat. However, Sirieix and Tagbata [5] showed that the ethical concept alone is not sufficient for organic consumers, who also demand information about the characteristics and quality of the product. These issues, together with consumers' direct experience with organic beef, play a major role in the repurchase decision [6] and thus avoid the discord between the expected and experienced quality [7].

Previous studies have been focused on factors regarding organic production systems, health issues, animal welfare, and the marketing of organic products [8–10]. Research on the quality characteristics of organic food products has mainly focused on animal nutrition aspects and its impact on the quality of the products [9], as well as the effects of organic food on human health [10,11].

In the case of organic meat, several studies indicated a higher content of ω-3 fatty acids in organic meat [9] as well as a higher nutritional value and bioactive compounds content in organic beef [12,13]. On the other hand, the physical–chemical quality aspects involved in the decision to buy and repurchase organic meat, such as colour, water content, and oxidative status as well as their conservation in different types of packaging [11,13,14], have been less studied. Despite these studies, it is necessary to increase research in order to improve the knowledge on the attributes of organic meat to strengthen the organic meat quality brand [15].

In this sense, the tenderness is the most valued quality attribute by consumers in beef [8,15]; consumers are often prepared to pay a higher price if the tenderness of the beef is guaranteed [15,16]. However, this is a complex parameter that results from the combination of several factors, such as animal rearing, intrinsic characteristics such as breed, gender, muscle structure, and connective tissue content, or the histological characteristics of muscle fibres, mainly sarcomere length and muscle fibre biochemistry [17]. The effect of the production system on the texture of beef has been studied by different authors, who described that the meat from calves reared in grazing production systems is tougher than animals reared under conventional feed-lot systems [18–20]. However, the literature about organic beef and production system effects on the texture properties of meat is scarcer [14,19]. The ageing process provides substantial improvements in meat tenderness and has been extensively studied [21–24]. This process involves complex changes in the muscle metabolism at the post-slaughter period, which can vary depending on animal breed, the metabolic status, and extrinsic conditions, such as the rearing system and stress prior to slaughter [24,25].

In Southwestern Europe, traditional cattle production is based on the extensive grazing of rustic breed calves, such as the Retinta or Avileña breeds on Mediterranean rangelands known as *Dehesas*. The *Dehesa* is a managed, agrosilvopastoral ecosystem whose soils are acidic, shallow, and easily eroded. Holm and cork oak form the most representative tree cover. The climate is semi-arid Mediterranean. The pasture that grows in these soil and climate conditions comprises many species, most of which are annuals. Cattle form the predominant livestock, with breeds of high rusticity that are able to thrive in this difficult environment. According to Caballero and Mata [26], this system is easily convertible to organic production by adjusting farm's health and management practices to the specifications of the EU's organic regulation [27]. Extensive conditions often involve periods of pasture shortage due to a dependence on climate conditions [28], which is a problem for producers. Although the term "organic" is often associated with free-range livestock [29], organic farmers have an alternative way to raise organic cattle without depending on the seasonal availability of grass [8] through the use of organic concentrate. Retinta is a rustic bovine breed that is well adapted to the *Dehesa* ecosystem, and it produces meat of high sensorial quality [30], with a lower intramuscular fat content [31] and a higher content of polyunsaturated fatty acids than other improved breeds [32]. However, in the meat industry market, where the meat yield of the carcass is a quality characteristic, carcasses of the Retinta breed are disadvantaged compared to improved breeds of cattle. Among the strategies to improve their value on the market, and considering their rustic character, organic production could give an added value to Retinta calves.

Within this framework, this paper attempts to raise knowledge about organic beef quality. The aim of this research is to analyse the effects of two types of organic production systems (grass-fed and organic concentrate-fed) by comparing both types with conventionally produced Retinta beef, and the effect of ageing period (0, 7, 14, and 21 days) on texture properties of beef from the Retinta breed.

2. Materials and Methods

2.1. Experimental Design and Animal Management

For the present study carried out in three experimental farms in southwestern Spain, seventy-five Retinta male calves were selected and reared with their mothers' milk until weaning at 8–9 months of age. At weaning, animals were allocated to three experimental groups ensuring that the average weight of each group was similar and where the calves were maintained since weaning until slaughter, as follows: organic grass-fed animals (OG, $n = 30$), organic concentrate-fed animals (OC, $n = 30$), and conventional concentrate-fed animals (CC, $n = 15$).

Calves in the OG system were fattened in a local agro-silvo-pastoral system called *Dehesa* in the experimental farm owned by CICYTEX (Center for Scientific and Technological Research of Extremadura, Spain). The calves were free grazing fed (from weaning to slaughter) on natural pasture resources from *Dehesa*, mainly composed of raygrass (*Lolium perenne* and *Lolium rigidum*) and clover (*Trifolium repens*), and they also received organic concentrate in controlled feeders when grass pasture was scarce (approximately 20% of the total dry matter supplied) according to Organic European Regulations [27]. The animals had freedom of movement and free access to water in the natural resources and waterers. Calves in the OC system were reared in pens allowing 8 m^2 per animal, according to regulations on organic production [27] at the Divino Salvador Coop. farm (Cádiz, Spain). After weaning, the animals were fed on a 40% organic concentrate and 60% organic forage—25% barley straw and 35% grass silage—(dry matter basis). The organic concentrate was composed of barley grain (36.2%); oat grain (24.5%); peas (16.6%); sunflower seed cake (19.6%); and minerals and vitamins (3.1%). The CC system animals were assigned to this research by the Diputación de Cádiz Agricultural Station (Cádiz, Spain). The calves were confined in pens allowing 4 m^2 per animal and fed on conventional concentrate *ad libitum* (approximately 80% of total dry matter supplied) and barley straw without the possibility of grazing. The conventional concentrate was composed of maize grain (34.0%); barley grain (33.5%); corn gluten feed (17.1%); soybean meal 44 (8.4%); minerals and vitamins (3.9%); and palm oil (3.1%).

The experimental procedures to which the animals were subjected during the fattening phase were in compliance; in the case of conventional production, they were considered standard farming practices and exempted from the consideration of ethical and welfare aspects by the Animal Care and Ethics Committee (CAEC). In the case of organic productions, animals were complying with animal welfare and organic regulations [27]. For all animals, Council Regulation (EC) No. 1099/2009 [33], for the protection of animals at the time of slaughter, was also complied.

Animals were slaughtered in 2 years (15 calves each year from each group) at their commercial weight in the local market in licensed slaughterhouses, which complied with animal welfare and organic regulations [27].

2.2. Muscle Sampling and Ageing Process

At 24h post-mortem, the pH$_{24}$ was measured on the *Longissimus thoracis* (LT) muscle at the 6th rib each of the left half carcass, and after the LT muscles were removed between the 5th and 10th rib. The *Longissimus thoracis* muscles were filleted from the cranial to caudal area, and eight steaks were obtained. The first four consecutive steaks (2 cm thick chops) were assigned to determine pH, drip loss, and histological analysis, one for each studied ageing time, and the next four steaks (3.5 cm thick chops) were assigned to texture study by post-mortem ageing time. The first steak was immediately analysed (T$_0$), and the fifth one was vacuum-packed in a plastic bag polyethylene (O$_2$ permeability, 9.3 mL O$_2$/m^2/24 h at 0 °C) and frozen at −20 °C for texture analysis. The steaks T$_0$ were analysed at 24 h post-mortem.

For the ageing process, the six remaining steaks were preserved individually, overwrapped with transparent oxygen-permeable polyvinyl chloride film for 7 (T$_7$), 14 (T$_{14}$), and 21 (T$_{21}$) days at 2–4 ± 1 °C in a refrigerator (mod. AN1002, Infrico S.L., Sevilla, Spain). After each ageing time, the samples for pH, drip loss, and histological study were analysed immediately, and the steaks for texture were vacuum-packed (under the above conditions) and frozen at −20 °C until analysis.

2.3. Physical–Chemical Analysis

The pH values were measured using a penetration electrode coupled with a temperature probe (Crison pH-meter mod. 507. Crison Instruments, Alella, Barcelona, Spain). The pH assessment of the ageing samples (T_7, T_{14}, and T_{21}) were measured after each ageing period.

Water losses were measured as drip loss (DL) and cooking loss (CL). DL was determined according to the method proposed by Honikel and Hamm [34] by duplicates on 50 g fresh samples taken and placed in a container (meat juice collector, Sarstedt, Nümbrecht, Germany) and kept at 4 ± 1 °C during the described ageing times. The results were expressed as water loss g/100 g of muscle between T_0 and T_7, T_7 and T_{14}, and T_{14} and T_{21}. The cooking loss method is described below, in the section about the instrumental texture.

The texture of meat was instrumentally evaluated on cooked samples. Previously, the samples were thawed in cold water (at room temperature) for 3 h before testing until reaching an internal temperature of 17–20 °C. For the cooking process, the raw samples were weighed and vacuum-packed in nylon/polyethylene bags and cooked by immersion in a water bath preheated at 80 °C, with controlled temperature until the steak reached an internal temperature of 75 °C [35]. Cooked samples were left to cool under tap water, to prevent further cooking, for 30 min and then chilled overnight at 4 °C. The difference in weight before and after cooking was used to calculate for the determination of cooking losses (CLs), and the results were expressed as water loss g/100 g of muscle.

For the texture analysis, each steak was cut transversally into two halves to be used in a Warner-Bratzler (WB) device and subject to texture profile analysis (TPA) with a compression device. For texture assessment, 1 cm^2 strips were made from each cooked steak, with the muscle fibers parallel to the longitudinal axis of the sample [36]. All texture measurements were taken using a texturometer TA-XT 2i Texture Analyser of Aname (Stable Micro Systems Ltd., Surrey, UK), and they were carried out with the sample at room temperature (22 ± 2 °C). Instrumental determinations were repeated 8 times per sample, and results were data averaged.

For WB analysis, the samples were cut with a Warner-Bratzler blade (HDP/BS) in the perpendicular direction to the muscle fibers. Three parameters were measured: maximum shear force (kg/cm^2), shear firmness (kg·s), and total work (kg/s).

On the other hand, the TPA test was conducted to evaluate the textural profile of the meat according to Tejerina et al. [37]. The cooked samples were cut into uniform cubes of approximately 1 cm^3 and were axially compressed to 20% (TPA20) of their original height using a probe with a 20 mm diameter flat plunger (P/20) connected to a load cell of 25 kg at a test speed of 2 mm/s. The samples were compressed in two cycle sequences, according to the recommendations for analysing food texture provided by Bourne [38]. TPA20 was used to determine the contribution of myofibrillar structures, without the intervention of connective tissue, on meat texture. The textural parameters obtained from force–deformation curves [39] were as follows: hardness (kg/cm^2) = maximum force required to compress the sample (peak force during the 1st compression cycle); springiness (cm) = height that the sample recovers during the time that elapses between the end of the 1st compression and the start of the 2nd; chewiness (kg·cm·s) = the work needed to chew a solid sample to a steady state of swallowing; and resilience (dimensionless) = how well the product regains its original height, as measured on the first withdrawal of the cylinder.

2.4. Histological Analysis of Muscle

Histological analyses (sarcomere length and cross-sectional area of fibre) of LT muscle were performed. Samples were taken directly from the centre of chops of the LT muscle collected and aged (for 0, 7, 14, and 21 days at 4 °C) and were placed in glutaraldehyde (2.5% *v/v* in phosphate buffer pH 6.5). The method used to determine sarcomere length was described by Torrescano et al. [40].

From each of the samples in glutaraldehyde, 4 bundles were removed under a magnifying glass and placed on glass slides, after which they were contrast-stained with haematoxylin and eosin. Sarcomere lengths were measured using an immersion objective (×100) under a phase contrast microscope (Nikon Eclipse 50i model) and Nis-Elements 3.10 computer image analysis software. The result was the average of 150 measured lengths (µm). Serial cross-sections were obtained according to described method by Abreu et al. [41]. Each of the LT muscle samples were frozen in liquid nitrogen and embedded in Tissue-Tek (Sakura Finetek Europe, Zoeterwoude, The Netherlands). Subsequently, serial cross-sections (10 µm thick) were cut with a cryostat at −20 °C and placed on poly-L-lysine-coated glass slides (Sigma-Aldrich, St. Louis, MO, USA), and they were stained for 30 s in haematoxylin. Between 100 and 150 cross-sections of fibre, randomly selected, were analysed per sample of LT from different production systems and ageing treatments. To determine the mean transverse of the muscular fibre (μm^2) by computerised image analysis, the Nis-Elements 3.10 software with a magnification of 100× was used. Structural elements were measured in a fibre bundle area, and more than 200 sections of samples were analysed per sample of LT from different ageing treatments and production systems.

2.5. Statistical Analysis

A two-way ANOVA test was carried out to determine the statistical significance for the 3 × 4 factorial design. The model included the fixed effects of the production system (OG, OC, and CC) and ageing time (D_0, D_7, D_{14}, and D_{21}) and their interactions on pH, water losses, instrumental textural properties, and histological parameters of LT muscle samples. Since there were significant differences in the slaughter weight among production systems, the effect of slaughter weight was used as a covariate. Slaughter weight was not included in the final model, because it did not have a significant influence on the parameters under study. The HSD Tukey's test was used to compare means, with significance being set at $p \leq 0.05$. Mean values and standard errors of the means (SEM) for all studied variables were reported. The relationship between the histological analysis variables of muscle fibre (sarcomere length and cross-sectional area of fibre) and texture parameters (Warner-Bratzler shear force and Texture Profile Analysis) were calculated by Pearson correlation coefficient (r).

3. Results and Discussion

3.1. pH and Water Losses (Drip and Cooking Losses)

The pH values of the LT muscle are shown in Table 1. An interaction between production system and ageing time was observed ($p \leq 0.05$) on pH. Consequently, the evolution of the pH throughout the ageing period was determined by the pH_{24} of the meat in each production system. Organic beef from the OG and OC production systems showed a lower pH than meat from the CC system. These values were below 6.0, which is in the normal range for beef [42], and therefore, it is without consequences from the point of view of meat quality. Our results are in accordance with those observed by Avilés et al. [43] for the Retinta breed. Regarding the production system, our findings are in line with other authors [25,44] who observed a lower pH value in beef from extensive production than in confined cattle. However, since the post-mortem decline in pH is related to muscle glycogen reserves, for confined cattle, which are accustomed to handling and contact with people, a lower pH due to a lower consumption of muscle glycogen was expected because of pre-slaughter stress [25]. Thus, any factor that increases the glycogen depletion rate leads to a decrease in substrate for post-mortem anaerobic glycolysis [45,46] and the consequent higher pH. The pH_{24} determines the quality characteristics of the meat. Specifically, Jeleníková et al. [47] concluded that there was a curvilinear relationship between the tenderness of the *Longissimus lumborum et thoracis* muscle and the final pH. Concerning ageing time, a significant effect ($p \leq 0.05$) on the pH values of the LT muscle was observed, which was also reported by Franco et al. [48].

Table 1. Effect of the production systems (organic grazing, OG; organic concentrate, OC; and conventional concentrate, CC) and ageing time (T_0, T_7, T_{14}, T_{21}: 0, 7, 14, and 21 days, respectively) on pH, water-holding capacity, and textural instrumental parameters of *Longissimus thoracis* from calves of the Retinta breed.

	Production System (PS)			Ageing Time (A)					Effects		
	OG	OC	CC	T_0	T_7	T_{14}	T_{21}	SEM	PS	A	PS*A
pH	5.55 [b]	5.54 [b]	5.60 [a]	5.52 [b]	5.56 [a,b]	5.60 [a,b]	5.63 [a]	0.017	*	*	*
Drip Loss (g/100 g)	5.42 [b]	4.44 [b]	6.80 [a]	-	3.04 [c]	4.28 [b]	6.48 [a]	0.197	***	***	***
Cooking Loss (g/100 g)	24.48 [a]	25.36 [a]	22.25 [b]	25.61 [b]	27.51 [a]	24.97 [b,c]	23.64 [c]	0.217	***	***	***
Warner-Bratzler shear force test											
WB-Shear Force (kg/cm^2)	6.79 [a]	5.21 [b]	5.08 [b]	8.66 [a]	6.67 [b]	5. 04 [c]	4.32 [c]	0.171	***	***	***
WB-Shear Firmness (kg/s)	11.23 [a]	8.15 [b]	8.76 [a,b]	13.07 [a]	9.82 [b]	8.64 [b,c]	7.37 [c]	0.25	***	***	***
WB-Total Work (kg*s)	1.30 [a]	1.15 [b]	1.07 [b]	1.71 [a]	1.27 [b]	1.02 [c]	0.90 [c]	0.03	***	***	***

* $p \leq 0.05$; *** $p \leq 0.001$; PS: production systems; A: ageing time; PS*A interaction between A and PS; SEM.: standard error of the mean. Values with the same letters (a, b, c) indicate homogeneous subsets for $p = 0.05$ according to Tukey's HSD test; OG: organic grazing; OC: organic concentrate; CC: conventional concentrate.

Table 1 shows the results of the effect of the production systems and ageing period on water losses (DL and CL). The water losses were affected by the significant interaction between the production system and ageing time factors ($p \leq 0.001$). These findings indicated that the effect of ageing time should be assessed within each production system.

Differences between meat produced under organic production systems (OG and OC) and meat produced under CC systems were observed, which showed the highest degree of DL in CC. Concerning the ageing time, the highest water loss was observed between T_{14} and T_{21} ($p \leq 0.05$). Drip loss is as assessment of the loss of fluid from beef cuts due to the shrinkage of muscle proteins (actin and myosin) in the form of drip. Several factors influence it, such as the breed [49–51] and the production system [44,52]. The results of water loss measured as CL (Table 1) showed differences due to the production system effect. The findings showed the opposite in relation to DL values, as Olsson et al. [53] reported for organic pork. Thus, the meat from organic production systems (OG and OC) showed higher CL than the meat from CC. Although the water content in meat plays an important role in the perception of juiciness and tenderness by consumers, to our knowledge, the scientific literature is scarce about assessing water losses between organic versus conventional beef. Miotello et al. [14] observed no differences on cooking water loss between organic and conventional beef. Ribas-Agustí et al. [12] observed higher water content measured by using a near-infrared meat analyser, but controversially, they found no difference with conventional beef in a meta-analysis. On the other hand, Walshe et al. [13] evaluated the total water content (moisture content) and did not observe differences due to the organic versus conventional production system. The ageing process affected cooking water loss; thus, the highest CL values were observed after 7 days of ageing (T_7) and the lowest value in T_{21}. However, Straadt et al. [54] reported that cooking water loss increased up to 4 days after slaughter, and Palka [55] observed that the cooking losses were lower in meat when the ageing period increased.

While drip loss occurs due to the degradation of the cytoskeletal protein during ageing [56] and to the formation of drip channels due to the disappearance of the link between myofibrils, which is caused by enzymatic action [57], cooking water loss is due to the denaturation of muscle proteins caused by the effect of heat, which affects their structural characteristics [57]. Several studies indicated the relationship between the drop in pH and lower WHC due to the denaturation of proteins [57], thereby facilitating the formation of drip channels. However, in light of our findings, the lower water losses (drip loss) in meat from organic systems could be also explained by other factors such as the structural and metabolic muscular changes [25,58] produced by the physical exercise due to the greater availability of space in these systems (OG and OC) and a greater consumption of mainly α-tocopherol from the grass in their diet (results not shown), which has also been reported by other authors [12,59]. Such factors help avoid lipid oxidation and stabilise the cell membranes, thus delaying the loss of water [59].

3.2. Warner-Bratzler Shear Force Test (WB)

Table 1 shows the effect of the production system on instrumental texture parameters, as measured by the Warner-Bratzler test on cooked meat. The interactions between production systems and ageing time were significant in all studied WB parameters ($p \leq 0.001$); consequently, the interpretation of the main effects is complex, which suggested that they should be assessed within each production system. Although some authors [9,11] observed no differences on the shear force due to organic or non-organic production systems, our results indicated the importance of this effect. According to Vestergaard et al. [20], the higher level of physical activity on grazing, compared with the limited activity in intensive condition, is the main reason for the highest WB parameter values, and therefore a lower tenderness. Physical exercise is associated with the extensive production system [19], which involves a structural and biochemical change in the muscle fibres [60]. These reasons could explain our results obtained for OG meat in shear force parameters, which were probably due to the availability of space to do physical exercise alongside the grass-feeding. Some authors observed that the muscles undergo a series of physical and biochemical changes due to exercise-linked grass-feeding [18,19,61]; Jurie et al. [60] observed an increase in oxidative muscle fibres, which was mainly related to the grass-feeding system affecting quality aspects such as colour and texture.

However, other factors such as diet could be taken into account, especially when it contains a high level of fat and high marbling scores in the meat [62]. Although Giaretta et al. [63] observed a positive correlation between meat tenderness and fat content, in particular to intramuscular fat [64,65], other authors reported that the diet is probably not the major cause of tenderness differences [43]. These authors reported that the fatness scores of Retinta carcass showed lower subcutaneous fat thickness than other rustic breeds, and they had low intramuscular fat values (<2%) under different types of feeding, which could be a reason for thinking about the importance of exercise in the above diet.

Considering that there is a relationship between shear force and hardness as sensory attributes [66], our results, instrumentally obtained, are consistent with the assessment sensorial of Retinta organic beef carried out by consumers [67], in which they rated meat from organic grass-fed animals (OG) as the toughest.

The ageing period had an effect on WB parameters (Table 1). All the parameters (shear force, shear firmness, and total work) decreased over ageing time ($p \leq 0.001$), as observed by other authors [19,51]. At T_0, the highest WB parameters values were observed, and in both shear force and total work, values reduced from T_0 to T_{14}, although after 14 days of ageing, no changes were observed in these parameters. In the present study, shear force values at 14 days of ageing were higher than those reported by other authors [51,68] for conventional meat from the Retinta breed.

Figure 1 shows the different tenderisation patterns among OG, OC, and CC meats throughout the ageing process. Although at the end of the ageing period (T_{21}) the meat in the three production systems showed no differences in shear force values, the reduction in shear force at 7 days of ageing was 42.65% for OG, 36.22% for OC, and 39.83% for CC, and after the first week, no significant differences among the production systems were shown; therefore, differences in the tenderisation pattern were identified. Thus, while the tenderisation speed of OC and CC meat was similar, OG meat revealed a totally different pattern of tenderisation. Initially, it was the toughest, and it had a quicker tenderisation during the first days of ageing. This finding leads us to consider that the differences found are mainly related to the extensiveness of the OG production system, in which greater exercise affects the myofibrillar structure and characteristics of muscle, causing an increase in meat tenderness after a long ageing period [19,60].

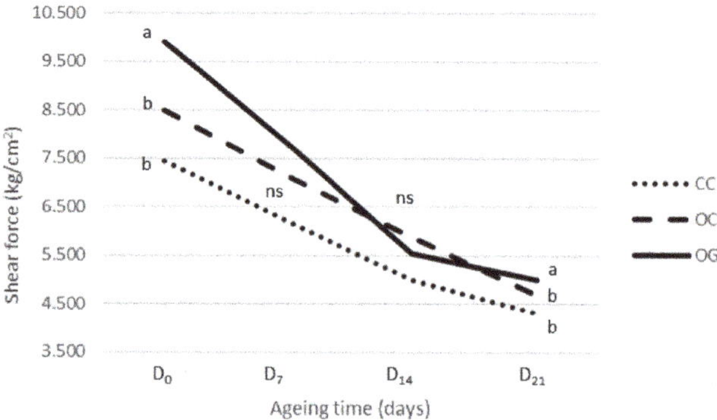

Figure 1. Evolution of the shear force values of *Longissimus thoracis* muscle from calves of the Retinta breed throughout 21 post-mortem days of ageing in organic grazing (OG), organic concentrate (OC), and conventional concentrate (CC) systems. Values with the same letters (a, b) indicate homogeneous subsets between productions systems for $p = 0.05$ according to Tukey's HSD test; ns = not significant.

3.3. Texture Profile Analysis (TPA)

Table 2 shows the results of effect of the production systems and ageing time on texture profile analysis at a compression ratio of 20% (TPA20). The compression parameters obtained with TPA have been used by many authors to effectively evaluate the textural properties of meat products in both raw and cooked meat [36,69,70]. The differences in the textural parameters of TPA20 are due to the behaviour of the myofibrillar structures without the intervention of the connective tissue [71].

Figure 2 showed that the production system had a significant interaction with ageing time for hardness ($p \leq 0.001$), springiness ($p \leq 0.01$), and resilience ($p \leq 0.05$), and the beef from each production system had different behaviour depending on the studied ageing time. Both the hardness and springiness parameters showed similar behaviours in organic meat (OG and OC), which were differentiated from conventional meat (CC).

Regarding hardness, the OG meat showed the highest values, the OC meat showed intermediate values, while the lowest values were found in the CC meat. The CC meat reached the lowest hardness value after 7 days of ageing, followed by the OC meat, which reached the lower value after 14 days, and finally the OG meat, which remained constant. On the other hand, the production system had a major weight on the interaction of the springiness, because ageing time showed no significant effect on this parameter. In the case of resilience, the highest value in OG meat was observed, and for both OC and CC, the lowest values were reached at 14 days of ageing. However, the chewiness and cohesiveness did not show significant interactions, and the main effects could be analysed separately. In both parameters, the production system effect results were significantly different ($p \leq 0.001$). The organic systems (OG and OC) showed the highest values of chewiness, and the OG meat showed the highest cohesiveness.

Thus, according to the definition of the parameters of Icier et al. [72], the organic beef (OG and OC) showed higher chewiness (length of time required to chew a sample to a consistency suitable for swallowing) and a lower springiness value (degree to which a product returns to its original shape once it has been compressed) than CC beef. In particular and in line with the one previously observed in shear force, the OG meat presented higher hardness (force necessary to attain a given deformation, maximum force) and cohesiveness values (strength of the internal bonds making up the body of the product).

Despite the difficulty of evaluating each factor linked to the production system individually (breed, gender, diet, available space, or exercise), due to the interaction among them, and in particular under extensive conditions, where grazing involves higher exercise, some studies suggest that the diet of the animals is not the most important factor in the tenderness of the meat [43]. Several studies had reported the exercise factor as determinant of meat texture characteristics. Thus, Aalhus and Price [73] concluded that exercise is a factor that can influence the type, size, and composition of muscle fibre and, therefore, this affects meat texture characteristics. The same finding was also reported by Jurie et al. [60], even when the space available was smaller. In light of our results, where each studied group had different available space, and according to the previous authors, we could highlight the exercise as a differentiating factor involved in the animals grazing.

Regarding the ageing time effect, it was noted that ageing time had a significant effect on chewiness ($p \leq 0.05$) parameters. As expected, at D_0, the hardness parameter showed the highest value, and the lowest value was observed in the chewiness parameter.

The effect of ageing time on texture characteristics is described as an effect that is more important than breed [22] and other factors.

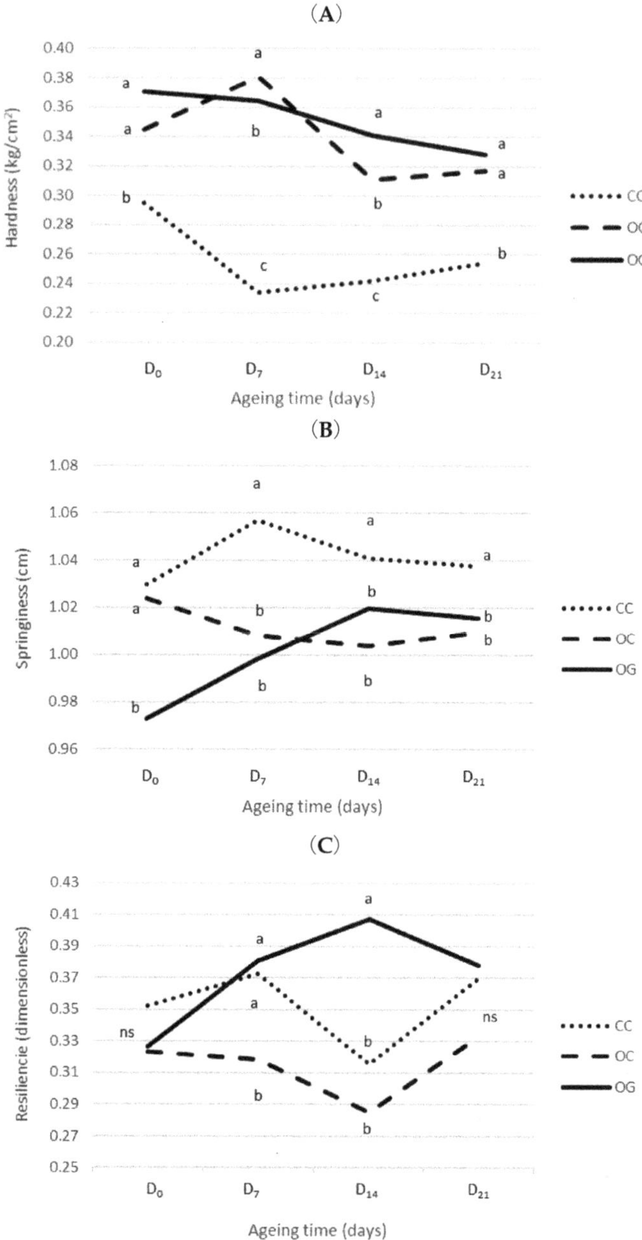

Figure 2. Interaction effects between production system and ageing time on TPA20 of *Longissimus thoracis* muscle from calves of the Retinta breed throughout 21 post-mortem days of ageing on Hardness (**A**), Springiness (**B**) and Resilence (**C**). Values with the same letters (a, b, c) indicate homogeneous subsets for $p = 0.05$ between productions systems according to Tukey's HSD test; ns = not significant; OG: organic grazing; OC: organic concentrate; CC: conventional concentrate.

Table 2. Effect of the production systems and ageing time (T_0, T_7, T_{14}, and T_{21}: 0, 7, 14, and 21 days, respectively) on texture profile analysis 20% (TPA20) of *Longissimus thoracis* from calves of the Retinta breed.

	Production System (PS)			Ageing Time (A)					Effects		
	OG	OC	CC	T_0	T_7	T_{14}	T_{21}	SEM	PS	A	PS*A
Hardness (kg/cm^2)	0.51 [a]	0.33 [b]	0.25 [c]	0.39 [a]	0.41 [a]	0.35 [b]	0.36 [b]	0.010	***	***	***
Springiness (cm)	1.00 [b]	1.01 [b]	1.04 [a]	1.00	1.02	1.01	1.02	0.003	***	ns	**
Resilience (non-dimensional)	0.37 [a]	0.15 [b]	0.35 [b]	0.35	0.36	0.35	0.34	0.003	***	ns	*
Chewiness (kg·m·s^{-2})	0.34 [a]	0.32 [a]	0.28 [b]	0.30 [b]	0.31 [a]	0.27 [a,b]	0.24 [a,b]	0.005	***	*	ns
Cohesiveness (non-dimensional)	0.65 [a]	0.62 [b]	0.62 [b]	0.64	0.64	0.63	0.63	0.003	***	ns	ns

ns: $p > 0.05$; ** $p \leq 0.01$; *** $p \leq 0.001$; A: ageing time; PS: production systems; PS*A: interaction between A and PS; SEM.: standard error of the mean. Values with the same letters (a, b, c, d) indicate homogeneous subsets for $p = 0.05$ according to Tukey's HSD test. OG: organic grazing; OC: organic concentrate; CC: conventional concentrate.

3.4. Muscle Fibre Characteristics

Table 3 shows the results obtained for the histological parameters of LT muscle fibres in Retinta calves. These results are in agreement with those found by other authors for beef (2.2 μm) [74]. The relationship between the characteristics of the muscle fibres and the texture of the meat has been extensively studied, both the relationship with the length of the sarcomere [75] and with the area of the muscle fibre [76–78].

The production system had a significant effect on sarcomere length ($p \leq 0.001$) and muscle fibre cross-sectional area ($p \leq 0.001$). OG meat muscle fibre was found to have a shorter sarcomere and larger fibre cross-sectional area than OC and CC. On the other hand, OC meat showed intermediate values on cross-sectional area, while the lowest values were found in the CC meat. Muscle fibre structures can change due to several factors such as age or exercise [79,80], which supports our hypothesis that extensive systems have a great influence on texture properties. As other authors indicated [20, 60], the extensive production system—due mainly to physical exercise—affects the histological characteristics of muscle fibres.

The histological parameters analysed (sarcomere length and cross-sectional area of fibres) showed significant differences throughout ageing time (Table 3). An increase in sarcomere length during ageing time ($p \leq 0.001$) was observed. This result is according with the findings of Viera and Fernández [81], who reported that the length of sarcomeres increases with ageing time and is closely related to shear force, which is in line with the observations of Ertbjerg and Puolanne [82]. In our study, the lowest sarcomere length value was observed at T_0, and the highest was observed at T_{21}. The greatest elongation of the sarcomere between two ageing times was observed from T_7 to T_{14}, and this result is consistent with the T_{14} instrumental measurements of shear force, although Starkey et al. [83] suggested that there is still a variation in shear force that cannot be completely explained by histological parameters such as sarcomere length, even though it is moderately correlated with shear force. Concerning the cross-section area, during 21 days of ageing, a continuous reduction until around 20% of the fibre cross-sectional area was noted, with the greatest decrease occurring between days T_0 and T_7.

Table 3. Effect of the production systems (organic grazing, OG; organic concentrate, OC and conventional concentrate, CC) and ageing time (T_0, T_7, T_{14}, T_{21}: 0, 7, 14, and 21 days, respectively) on histological parameters of *Longissimus thoracis* from calves of the Retinta breed.

	Production System (PS)			Ageing Time (A)					Effects		
	OG	OC	CC	T_0	T_7	T_{14}	T_{21}	SEM	PS	A	PS*A
Sarcomere length (μm)	2.31 [b]	2.89 [a]	2.98 [a]	1.92 [d]	2.14 [c]	3.02 [b]	3.40 [a]	0.026	***	***	***
Cross-section Area (μm²)	1291.24 [a]	1177.75 [b]	921.61 [c]	1217.27 [a]	1108.32 [b]	1035.07 [b,c]	978.33 [c]	13.525	***	***	***

*** $p \leq 0.001$; A: ageing time; PS: production systems; PS*A: interaction between A and PS; SEM. standard error of the mean. Values with the same letters (a, b, c, d) indicate homogeneous subsets for $p = 0.05$ according to Tukey's HSD test. OG: organic grazing; OC: organic concentrate; CC: conventional concentrate.

3.5. Relationship between Texture Parameters and Muscle Fibre Characteristics

Table 4 shows the correlation coefficients between histological and instrumental texture variables. Our data confirm a negative correlation between cross-sectional area and sarcomere length (-0.284; $p \leq 0.01$). A negative correlation between sarcomere length (-0.355; $p \leq 0.01$) and the shear force was also observed. These results agree with those reported by Janz et al. [84], which established a correspondence between a longer sarcomere length and a significantly lower shear force value. Rhee et al. [75] also found that the overall sarcomere length was significantly correlated to tenderness, and Pen et al. [85] reported that sarcomere length plays an important role in meat tenderness, particularly in the first few days of ageing until proteolysis exerts a greater influence than sarcomere length on the final texture of the meat, since the ageing of beef is considered the most important factor myofibrillar tenderisation [22]. On the other hand, Żochowska et al. [77] and Guillemin et al. [78] informed that meat with higher fibre diameters had greater shear strength, although ours findings showed no correlation ($p > 0.05$).

Table 4. Pearson's correlation coefficient (r) between histological and texture parameters of *Longissimus thoracis* from calves of the Retinta breed.

	Sarcomere	Area	Shear F.	Hard20	Spri20	Chew20	Res20	Coh20
Sarcomere	1							
Area	−0.284 **	1						
Shear F.	−0.355 **	−0.080	1					
Hard20	−0.444 **	0.261 **	−0.192	1				
Spri20	0.372 **	−0.215 *	0.152	−0.956 **	1			
Chew20	0.362 **	−0.162	0.393 **	−0.382 **	0.475 **	1		
Res20	−0.183 *	0.110	−0.062	0.847 **	−0.866 **	−0.163 *	1	
Coh20	−0.415 **	0.254 **	−0.264	0.985 **	-0.947 **	-0.394 **	0.872 **	1

* $p \leq 0.05$; ** $p \leq 0.01$; Sarcomere: sarcomere length; Area: cross-section area; Shear F: shear force; compression test TPA at 20% deformation: Hard20 (hardness); Spri20 (springiness); Chew20 (chewiness); Res20 (resilience); Coh20 (cohesiveness).

Sarcomere length was negatively correlated with some TPA20 parameters as hardness (-0.444; $p \leq 0.01$), resilience (-0.180; $p \leq 0.05$), and cohesiveness (-0.415; $p \leq 0.01$), and it was correlated positively with chewiness (0.372; $p \leq 0.01$) and springiness (0.362; $p \leq 0.01$). On the other hand, cross-section area values showed positive correlations on hardness (0.261; $p \leq 0.01$) and cohesiveness (0.254; $p \leq 0.01$), whilst the correlation was negative with springiness (-0.215; $p \leq 0.05$).

A lack of correlation between Warner-Bratzler and compression test (TPA) parameters is usually found [68], as noted in this study. In this sense, our results only indicated a positive correlation between shear force and chewiness (0.393; $p \leq 0.01$). Caine et al. [86] observed a similar value for this correlation, and they suggested that both instrumental methods (Warner-Bratzler and TPA tests) could be useful in assessing meat hardness. Thus, TPA could be a valid method for the evaluation of meat texture characteristics [36].

4. Conclusions

OG production systems resulted in meat with higher instrumentally measured textural values (Warner-Bratzler and TPA at 20% compression). Overall, the meat was harder, with less springiness, more resilient, and more cohesive than OC and CC meat samples. In spite of these characteristics, the OG meat pattern of tenderness, i.e., the fast decrease of shear force values throughout its ageing time, was greater in the first days of ageing when compared to the ones from the other systems studied. After 14 days, meat tenderness is similar in the three systems, and it can now be established that 14 days is the ageing time needed for the Retinta breed meat to reach its maximum value of tenderness. As a consequence, extending the ageing time does not improve meat tenderness; however, it could result in an increase of production costs. Furthermore, and in line with instrumental texture results, our findings show a direct association between muscle fibre characteristics, which showed changes throughout the ageing period, and meat tenderness from the different production systems. Thus, the sarcomere's length was directly proportional to those parameters related to tenderness (shear force or hardness), while the cross-sectional area of muscle fibres was inversely proportional to them. However, more biochemically or histochemically oriented research is needed to better understand the effect of differentiating factors of the production systems (available space or exercise) on textural beef characteristics.

Given our results, Retinta organic meat is a more suitable choice for consumers who, compared to a conventional diet, perceive organic food as more valuable, since tenderness has been associated to meat repurchase consumer behaviour. Thus, this positive consumer experience of organic Retinta meat would contribute to consumer loyalty and improve the development of organic meat production.

Author Contributions: S.G.-T. and A.H. conceived the idea, funding acquisition, writing—original draft and participated with all authors. A.L.-G. and E.P. elaborated the samples and analysed it. M.C.d.V. and D.T. responsible of methodology and elaborated formal analysis. All authors have read and agreed to the published version of the manuscript.

Funding: This paper received the support of the Spanish National Institute for Agriculture, Food Research and Technology (INIA) project (RTA2009-00122-C03-00).

Acknowledgments: Dª Adoración López Gajardo would like to thank the INIA for a PhD grant. The authors would like to thank the staff of farm Divino Salvador Cooperative (Cádiz), the Diputación of Cádiz and CICYTEX "Finca Valdesequera" (Badajoz) for their technical support and gratefully acknowledge the help and advice of the technical staff at Laboratory of Meat Quality, especially of Catalina Osorio and María Pérez.

Conflicts of Interest: The authors declare no conflict of interest.

References

1. Jensen, K.O.D.; Denver, S.; Zanoli, R. Actual and potential development of consumer demand on the organic food market in Europe. *NJAS—Wagening. J. Life Sci.* **2011**, *58*, 79–84. [CrossRef]
2. Lee, H.J.; Yun, Z.S. Consumers' perceptions of organic food attributes and cognitive and affective attitudes as determinants of their purchase intentions toward organic food. *Food Qual. Prefer.* **2015**, *39*, 259–267. [CrossRef]
3. Gil, M.; Gracia, J.M.; Sánchez, A. Market segmentation and willingness to pay for organic products in Spain. *Int. Food Agribus. Manag. Rev.* **2000**, *3*, 207–226. [CrossRef]
4. Napolitano, F.; Braghieri, A.; Piasentier, E.; Favotto, S.; Naspetti, S.; Zanoli, R. Effect of information about organic production on beef liking and consumer willingness to pay. *Food Qual. Prefer.* **2010**, *21*, 207–212. [CrossRef]
5. Sirieix, L.; Tagbata, D. Consumers willingness to pay for fair trade and organic products. In *Cultivating the Future Based on Science, Proceedings of the 2nd Scientific Conference of the International Society of Organic Agriculture Research, Modena, Italy, 18–20 June 2008*; pp. 260–263.
6. Banović, M.; Grunert, K.G.; Barreira, M.M.; Fontes, M.A. Beef quality perception at the point of purchase: A study from Portugal. *Food Qual. Prefer.* **2009**, *20*, 335–342. [CrossRef]
7. Henchion, M.M.; McCarthy, M.; Resconi, V.C. Beef quality attributes: A systematic review of consumer perspectives. *Meat Sci.* **2017**, *128*, 1–7. [CrossRef]
8. Gajardo, A.L. Parámetros de Calidad y Características Sensoriales de la Carne de Terneros de Raza Retinta Criados en dos Modelos de Producción Ecológica. Ph.D. Thesis, Universidad de Sevilla, Seville, Spain, 2018.
9. Bjorklund, E.A.; Heins, B.J.; DiCostanzo, A.; Chester-Jones, H. Fatty acid profiles, meat quality, and sensory attributes of organic versus conventional dairy beef steers. *J. Dairy Sci.* **2014**, *97*, 1828–1834. [CrossRef]
10. Popa, M.E.; Mitelut, A.C.; Popa, E.E.; Stan, A.; Popa, V.I. Organic foods contribution to nutritional quality and value. *Trends Food Sci. Technol.* **2019**, *84*, 15–18. [CrossRef]
11. Kamihiro, S.; Stergiadis, S.; Leifert, C.; Eyre, M.D.; Butler, G. Meat quality and health implications of organic and conventional beef production. *Meat Sci.* **2015**, *100*, 306–318. [CrossRef]
12. Ribas-Agustí, A.; Díaz, I.; Sárraga, C.; García-Regueiro, J.A.; Castellari, M. Nutritional properties of organic and conventional beef meat at retail. *J. Sci. Food Agric.* **2019**, *99*, 4218–4225. [CrossRef]

13. Walshe, B.E.; Sheehan, E.M.; Delahunty, C.M.; Morrissey, P.A.; Kerry, J.P. Composition, sensory and shelf life stability analyses of Longissimus dorsi muscle from steers reared under organic and conventional production systems. *Meat Sci.* **2006**, *73*, 319–325. [CrossRef] [PubMed]

14. Miotello, S.; Bondesan, V.; Tagliapietra, F.; Schiavon, S.; Bailoni, L. Meat quality of calves obtained from organic and conventional farming. *Ital. J. Anim. Sci.* **2009**, *8*, 213–215. [CrossRef]

15. Verbeke, W.; van Wezemael, L.; de Barcellos, M.D.; Ku, J.O. European beef consumers' interest in a beef eating-quality guarantee Insights from a qualitative study in four EU countries. *Appetite* **2010**, *54*, 289–296. [CrossRef] [PubMed]

16. Monsón, F.; Sañudo, C.; Sierra, I. Influence of breed and ageing time on the sensory meat quality and consumer acceptability in intensively reared beef. *Meat Sci.* **2005**, *71*, 471–479. [CrossRef]

17. Koohmaraie, M.; Geesink, G.H. Contribution of postmortem muscle biochemistry to the delivery of consistent meat quality with particular focus on the calpain system. *Meat Sci.* **2006**, *74*, 34–43. [CrossRef]

18. Nuernberg, K.; Dannenberger, D.; Nuernberg, G.; Ender, K.; Voigt, J.; Scollan, N.D.; Wood, J.D.; Nute, G.R.; Richardson, R.I. Effect of a grass-based and a concentrate feeding system on meat quality characteristics and fatty acid composition of longissimus muscle in different cattle breeds. *Livest. Prod. Sci.* **2005**, *94*, 137–147. [CrossRef]

19. Marino, R.; Albenzio, M.; Braghieri, A.; Muscio, A.; Sevi, A. Organic farming: Effects of forage to concentrate ratio and ageing time on meat quality of Podolian young bulls. *Livest. Sci.* **2006**, *102*, 42–50. [CrossRef]

20. Vestergaard, M.; Therkildsen, M.; Henckel, P.; Jensen, L.R.; Andersen, H.R.; Sejrsen, K. Influence of feeding intensity, grazing and finishing feeding on meat and eating quality of young bulls and the relationship between muscle fibre characteristics, fibre fragmentation and meat tenderness. *Meat Sci.* **2000**, *54*, 187–195. [CrossRef]

21. Sañudo, C.; Macie, E.S.; Olleta, J.L.; Villarroel, M.; Panea, B.; Albertí, P. The effects of slaughter weight, breed type and ageing time on beef meat quality using two different texture devices. *Meat Sci.* **2004**, *66*, 925–932. [CrossRef]

22. Campo, M.M.; Santolaria, P.; San, C. Assessment of breed type and ageing time effects on beef meat quality using two different texture devices. *Meat Sci.* **2000**, *55*, 371–378. [CrossRef]

23. Campo, M.M.; Sañudo, C.; Panea, B.; Alberti, P.; Santolaria, P. Breed type and ageing time effects on sensory characteristics of beef strip loin steaks. *Meat Sci.* **1999**, *51*, 383–390. [CrossRef]

24. Muchenje, V.; Hugo, A.; Dzama, K.; Chimonyo, M.; Strydom, P.E.; Raats, J.G. Cholesterol levels and fatty acid profiles of beef from three cattle breeds raised on natural pasture. *J. Food Compos. Anal.* **2009**, *22*, 354–358. [CrossRef]

25. Díaz, F.; Díaz-Luis, F.; Sierra, A.; Diñeiro, V.; González, Y.; García-Torres, P.; Oliván, S. What functional proteomic and biochemical analysis tell us about animal stress in beef? *J. Proteom.* **2020**, *218*, 103722. [CrossRef]

26. Caballero, C.; Mata, I. Posibilidades de la ganadería ecológica en Andalucía. In *Agricultura Ecológica y Desarrollo Rural, Proceedings of the II Congreso de la Sociedad Española de Agricultura Ecológica, Pamplona, Spain, September 1996*; pp. 117–127.

27. Council, E. Council regulation (EC) No 834/2007 of 28 June 2007 on organic production and labelling of organic products and repealing regulation (EEC) No 2092/91. *Off. J. Eur. Union* **2007**, *189*, 1–23.

28. García-Torres, S.; Díaz, M.E.; Parra, M.M.L.; Pérez, M.P.I.C.V. Sistemas de acabado de terneros en Extremadura. *ITEA* **1997**, *18*, 203–2015.

29. Mesías, F.J.; Pleite, F.M.; Martínez-Paz, J.M.; García, P.G. La disposición a pagar por alimentos ecológicos en España: Una aproximación a la existencia de diferencias regionales. *ITEA* **2011**, *107*, 3–20.

30. Alberti, M.M.; Sañudo, P.; Santolana, C.; Lahoz, P.; Olleta, F.; Campo, J.L. Características de la canal y calidad de la carne de añojos de la raza Retinta. *Arch. Zootec.* **1995**, *44*, 283.

31. Panea, B.; Ripoll, G.; Sañudo, C.; Olleta, J.L.; Albertí, P. Calidad instrumental de la carne de terneros procedentes del cruce industrial de la raza Retinta. *ITEA Inf. Tec. Econ. Agrar.* **2016**, *112*, 286–300.

32. Ibáñez, A.H.; López, A.; Polo, O.P.; Pino, R.; de la Vega, M.D.C.; Barrado, D.T.; Torres, S.G. Fatty acid profile as a tool to trace the origin of beef in pasture- and grain-fed young bulls of Retinta breed. *Span. J. Agric. Res.* **2017**, *15*, e0607.

33. Council Regulation, C. Council Regulation (EC) No. 1099/2009 of 24 September 2009 on the protection of animals at the time of killing. *Off. J. Eur. Union L* **2009**, *303*, 1.

34. Honikel, R.; Hamm, K.O. Measurement of water-holding capacity and juiciness. In *Quality Attributes and Their Measurement in Meat, Poultry and Fish Products*; Springer: Boston, MA, USA, 1994; pp. 125–161.

35. Combes, S.; Lepetit, J.; Darche, B.; Lebas, F. Effect of cooking temperature and cooking time on Warner-Bratzler tenderness measurement and collagen content in rabbit meat. *Meat Sci.* **2004**, *66*, 91–96. [CrossRef]

36. De Huidobro, F.R.; Miguel, E.; Blázquez, B.; Onega, E. A comparison between two methods (Warner-Bratzler and texture profile analysis) for testing either raw meat or cooked meat. *Meat Sci.* **2005**, *69*, 527–536. [CrossRef] [PubMed]

37. Tejerina, D.; García-Torres, S.; Cava, R. Water-holding capacity and instrumental texture properties of m. Longissimus dorsi and m. Serratus ventralis from Iberian pigs as affected by the production system. *Livest. Sci.* **2012**, *148*, 46–51. [CrossRef]

38. Bourne, M.C. Sensory methods of texture and viscosity measurement. In *Food Texture and Viscosity: Concept and Measurement*; Stewart, G.F., Schweigert, B.S., Hawthorn, J., Eds.; Academic Press: New York, NY, USA, 1982; pp. 247–279.

39. Bourne, M. Texture Profile Analysis. *Food Technol.* **1978**, *32*, 62–66.

40. Torrescano, G.; Sánchez-Escalante, A.; Giménez, B.; Roncalés, P.; Beltrán, J.A. Shear values of raw samples of 14 bovine muscles and their relation to muscle collagen characteristics. *Meat Sci.* **2003**, *64*, 85–91. [CrossRef]

41. Abreu, E.; Quiroz-Rothe, E.; Mayoral, A.I.; Vivo, J.M.; Robina, Á.; Guillén, M.T.; Agüera, E.; Rivero, J.L.L. Myosin heavy chain fibre types and fibre sizes in nuliparous and primiparous ovariectomized Iberian sows: Interaction with two alternative rearing systems during the fattening period. *Meat Sci.* **2006**, *74*, 359–372. [CrossRef]

42. Miranda-de la Lama, G.C.; Villarroel, M.; María, G.A. Livestock transport from the perspective of the pre-slaughter logistic chain: A review. *Meat Sci.* **2014**, *98*, 9–20. [CrossRef]

43. Avilés, C.; Martínez, A.L.; Domenech, V.; Peña, F. Effect of feeding system and breed on growth performance, and carcass and meat quality traits in two continental beef breeds. *Meat Sci.* **2015**, *107*, 94–103. [CrossRef]

44. Cozzi, G.; Brscic, M.; da Ronch, F.; Boukha, A.; Tenti, S.; Gottardo, F. Comparison of two feeding finishing treatments on production and quality of organic beef. *Ital. J. Anim. Sci.* **2009**, *9*, 404–409. [CrossRef]

45. Muchenje, V.; Dzama, K.; Chimonyo, M.; Strydom, P.E.; Raats, J.G. Relationship between pre-slaughter stress responsiveness and beef quality in three cattle breeds. *Meat Sci.* **2009**, *81*, 653–657. [CrossRef]

46. Warriss, P.D. The handling of cattle pre-slaughter and its effects on carcass and meat quality. *Appl. Anim. Behav. Sci.* **1990**, *28*, 171–186. [CrossRef]

47. Jeleníková, J.; Pipek, P.; Staruch, L. The influence of ante-mortem treatment on relationship between pH and tenderness of beef. *Meat Sci.* **2008**, *80*, 870–874.

48. Franco, D.; Bispo, D.; González, E.; Vázquez, L.; Moreno, J.A. Effect of finishing and ageing time on quality attributes 2 of loin from the meat of Holstein-Fresian cull cows. *Meat Sci.* **2009**, *83*, 484–491. [CrossRef] [PubMed]

49. Oliván, M.; Martínez, A.; Osoro, K.; Sañudo, C.; Panea, B.; Olleta, J.L.; Campo, M.M.; Oliver, M.A.; Serra, X.; Piedrafita, J.; et al. Effect of muscular hypertrophy on physico-chemical, biochemical and texture traits of meat from yearling bulls. *Meat Sci.* **2004**, *68*, 567–575. [CrossRef] [PubMed]

50. Aldai, A.I.; Murray, N.; Oliván, B.E.; Martínez, M.; Troy, A.; Osoro, D.J.; Nájera, K. The influence of breed and mh-genotype on carcass conformation, meat physico-chemical characteristics, and the fatty acid profile of muscle from yearling bulls. *Meat Sci.* **2006**, *72*, 486–495. [CrossRef] [PubMed]

51. López-Pedrouso, M.; Rodríguez-Vázquez, R.; Purriños, L.; Oliván, M.; García-Torres, S.; Sentandreu, M.Á.; Lorenzo, J.M.; Zapata, C.; Franco, D. Sensory and Physicochemical Analysis of Meat from Bovine Breeds in Different Livestock Production Systems, Pre-Slaughter Handling Conditions and Ageing Time. *Foods* **2020**, *9*, 176. [CrossRef] [PubMed]

52. Strydom, P.; Lühl, J.; Kahl, C.; Hoffman, L.C. Comparison of shear force tenderness, drip and cooking loss, and ultimate muscle pH of the loin muscle among grass-fed steers of four major beef crosses slaughtered in Namibia. *South Afr. J. Anim. Sci.* **2016**, *46*, 348–359. [CrossRef]

53. Olsson, V.; Andersson, K.; Hansson, I.; Lundström, K.L. Differences in meat quality between organically and conventionally produced pigs. *Meat Sci.* **2003**, *64*, 287–297. [CrossRef]

54. Straadt, I.K.; Rasmussen, M.; Andersen, H.J.; Bertram, H.C. Aging-induced changes in microstructure and water distribution in fresh and cooked pork in relation to water-holding capacity and cooking loss—A combined confocal laser scanning microscopy (CLSM) and low-field nuclear magnetic resonance relaxation study. *Meat Sci.* **2007**, *75*, 687–695. [CrossRef]

55. Palka, K. The influence of post-mortem ageing and roasting on the microstructure, texture and collagen solubility of bovine semitendinosus muscle. *Meat Sci.* **2003**, *64*, 191–198. [CrossRef]

56. Kristensen, L.; Purslow, P.P. The efect of ageing on the water-holding capacity of pork: Role of cytoskeletal proteins. *Meat Sci.* **2001**, *58*, 17–23. [CrossRef]

57. Hughes, J.M.; Oiseth, S.K.; Purslow, P.P.; Warner, R.D. A structural approach to understanding the interactions between colour, water-holding capacity and tenderness. *Meat Sci.* **2014**, *98*, 520–532. [CrossRef] [PubMed]

58. Bond, J.J.; Warner, R.D. Ion distribution and protein proteolysis affect water holding capacity of Longissimus thoracis et lumborum in meat of lamb subjected to antemortem exercise. *Meat Sci.* **2007**, *75*, 406–414. [CrossRef] [PubMed]

59. Horcada, A.; Polvillo, O.; González-Redondo, P.; López, A.; Tejerina, D.; García-Torres, S. Stability of fatty acid composition of intramuscular fat from pasture- and grain-fed young bulls during the first 7 d postmortem. *Arch. Anim. Breed.* **2020**, *63*, 45.

60. Jurie, C.; Ortigues-Marty, I.; Picard, B.; Micol, D.; Hocquette, J.F. The separate effects of the nature of diet and grazing mobility on metabolic potential of muscles from Charolais steers. *Livest. Sci.* **2006**, *104*, 182–192. [CrossRef]

61. Frylinck, L.; Strydom, P.E.; Webb, E.C.; Toit, E.D. Effect of South African beef production systems on post-mortem muscle energy status and meat quality. *Meat Sci.* **2013**, *93*, 827–837. [CrossRef]

62. Mordenti, A.L.; Brogna, N.; Canestrari, G.; Bonfante, E.; Eusebi, S.; Mammi, L.M.; Giaretta, E.; Formigoni, A. Effects of breed and different lipid dietary supplements on beef quality. *Anim. Sci. J.* **2019**, *90*, 619–627. [CrossRef]

63. Giaretta, A.; Mordenti, E.; Canestrari, A.L.; Brogna, G.; Palmonari, N.; Formigoni, A. Assessment of muscle Longissimus thoracis et lumborum marbling by image analysis and relationships between meat quality parameters. *PLoS ONE* **2018**, *13*, e0202535. [CrossRef]

64. Blanco, M.; Casasús, I.; Ripoll, G.; Albertí, P.; Panea, B.; Joy, M. Is meat quality of forage-fed steers comparable to the meat quality of conventional beef from concentrate-fed bulls? *J. Sci. Food Agric.* **2017**, *97*, 4943–4952. [CrossRef]

65. Ripoll, G.; Blanco, M.; Albertí, P.; Panea, B.; Joy, M.; Casasús, I. Effect of two Spanish breeds and diet on beef quality including consumer preferences. *J. Sci. Food Agric.* **2014**, *94*, 983–992. [CrossRef]

66. Starkey, C.P.; Geesink, G.H.; Collins, D.; Oddy, V.H.; Hopkins, D.L. Do sarcomere length, collagen content, pH, intramuscular fat and desmin degradation explain variation in the tenderness of three ovine muscles? *Meat Sci.* **2016**, *113*, 51–58. [CrossRef] [PubMed]

67. García-Torres, S.; López-Gajardo, A.; Mesías, F.J. Intensive vs. free-range organic beef. A preference study through consumer liking and conjoint analysis. *Meat Sci.* **2016**, *114*, 114–120. [CrossRef] [PubMed]

68. Panea, J.; Olleta, B.; Sañudo, J.L.; Campo, C.; Oliver, M.M.; Gispert, M.A.; Serra, M.; Renand, X.; Oliván, G.; Jabet, M.C.; et al. Effect of breed-production system on collagen, textural, and sensory traits of 10 European beef cattle breeds. *J. Texture Stud.* **2018**, *49*, 1–8. [CrossRef] [PubMed]

69. De Ávila, M.D.R.; Cambero, M.I.; Ordóñez, J.A.; de la Hoz, L.; Herrero, A.M. Rheological behaviour of commercial cooked meat products evaluated by tensile test and texture profile analysis (TPA). *Meat Sci.* **2014**, *98*, 310–315. [CrossRef] [PubMed]

70. Chang, H.; Wang, Q.; Xu, X.; Li, C.; Huang, M.; Zhou, G.; Dai, Y. Effect of heat-induced changes of connective tissue and collagen on meat texture properties of beef Semitendinosus muscle. *Int. J. Food Prop.* **2011**, *14*, 381–396. [CrossRef]

71. Tejerina, D.; García-Torres, S.; de Vaca, M.C.; Vázquez, F.M.; Cava, R. Effect of production system on physical-chemical, antioxidant and fatty acids composition of Longissimus dorsi and Serratus ventralis muscles from Iberian pig. *Food Chem.* **2012**, *133*, 293–299. [CrossRef] [PubMed]

72. Icier, F.; Izzetoglu, G.T.; Bozkurt, H.; Ober, A. Effects of ohmic thawing on histological and textural properties of beef cuts. *J. Food Eng.* **2010**, *99*, 360–365. [CrossRef]

73. Aalhus, J.L.; Price, M.A. Endurance-exercised growing sheep: I. Post-mortem and histological changes in skeletal muscles. *Meat Sci.* **1991**, *29*, 43–56. [CrossRef]

74. Herring, E.J.; Cassens, H.K.; Suess, R.G.; Brungardt, G.G.; Briskey, V.H. Tenderness and Associated Characteristics of Stretched and Contracted Bovine Muscles. *J. Food Sci.* **1967**, *32*, 317–323. [CrossRef]

75. Rhee, M.S.; Wheeler, T.L.; Shackelford, S.D.; Koohmaraie, M. Variation in palatability and biochemical traits within and among eleven beef muscles. *J. Anim. Sci.* **2004**, *82*, 534–550. [CrossRef]

76. Renand, G.; Picard, B.; Touraille, C.; Berge, P.; Lepetit, J. Relationships between muscle characteristics and meat quality traits of young Charolais bulls. *Meat Sci.* **2001**, *59*, 49–60. [CrossRef]

77. Zochowska, J.; Lachowicz, K.; Gajowiecki, L.; Sobczak, M.; Kotowicz, M.; Zych, A. Effects of carcass weight and muscle on texture, structure and myofibre characteristics of wild boar meat. *Meat Sci.* **2005**, *71*, 244–248. [CrossRef] [PubMed]

78. Guillemin, N.; Jurie, C.; Cassar-Malek, I.; Hocquette, J.F.; Renand, G.; Picard, B. Variations in the abundance of 24 protein biomarkers of beef tenderness according to muscle and animal type. *Animal* **2011**, *5*, 885–894. [CrossRef] [PubMed]

79. Pette, D.; Staron, R.S. Transitions of muscle fiber phenotypic profiles. *Histochem. Cell Biol.* **2001**, *115*, 359–372. [CrossRef]

80. Joo, S.T.; Kim, G.D.; Hwang, Y.H.; Ryu, Y.C. Control of fresh meat quality through manipulation of muscle fiber characteristics. *Meat Sci.* **2013**, *95*, 828–836. [CrossRef]

81. Vieira, C.; Fernández, A.M. Effect of ageing time on suckling lamb meat quality resulting from different carcass chilling regimes. *Meat Sci.* **2014**, *96*, 682–687. [CrossRef]

82. Ertbjerg, E.; Puolanne, P. Muscle structure, sarcomere length and influences on meat quality: A review. *Meat Sci.* **2017**, *132*, 139–152. [CrossRef]

83. Starkey, C.P.; Geesink, G.H.; Oddy, V.H.; Hopkins, D.L. Explaining the variation in lamb longissimus shear force across and within ageing periods using protein degradation, sarcomere length and collagen characteristics. *Meat Sci.* **2015**, *105*, 32–37. [CrossRef]

84. Janz, J.A.M.; Aalhus, J.L.; Dugan, M.E.R.; Price, M.A. A mapping method for the description of Warner-Bratzler shear force gradients in beef Longissimus thoracis et lumborum and Semitendinosus. *Meat Sci.* **2006**, *72*, 79–90. [CrossRef]

85. Pen, S.; Kim, Y.H.B.; Luc, G.; Young, O.A. Effect of pre rigor stretching on beef tenderness development. *Meat Sci.* **2012**, *92*, 681–686. [CrossRef]

86. Caine, W.R.; Aalhus, J.L.; Best, D.R.; Dugan, M.E.R.; Jeremiah, L.E. Relationship of texture profile analysis and Warner-Bratzler shear force with sensory characteristics of beef rib steaks. *Meat Sci.* **2003**, *64*, 333–339. [CrossRef]

Article

The Mode of Grass Supply to Dairy Cows Impacts on Fatty Acid and Antioxidant Profile of Milk

Senén De La Torre-Santos [1], Luis J. Royo [1], Adela Martínez-Fernández [1], Cristina Chocarro [2] and Fernando Vicente [1,*]

[1] Servicio Regional de Investigación y Desarrollo Agroalimentario (SERIDA), Carretera AS-267, PK. 19, Villaviciosa, 33300 Asturias, Spain; senendelatorre@gmail.com (S.D.L.T.-S.); ljroyo@serida.org (L.J.R.); admartinez@serida.org (A.M.-F.)

[2] ETSEA, Universitat de Lleida, C/Rovira Roure 177, 25198 Lleida, Spain; cristina.chocarro@udl.cat

* Correspondence: fvicente@serida.org

Received: 30 July 2020; Accepted: 4 September 2020; Published: 8 September 2020

Abstract: The optimization of milk production includes a rational use of forages, respect for the environment and offers the best quality to consumers. Milk production based on grass and forages produces healthier milk and it is widely spread throughout the Atlantic arc to maximize milk yield per hectare. However, the mode of offering the grass can have a major influence on milk composition. The aim of this study was to evaluate the effect of grass supply mode (grazing, zero-grazing or ensiling) on dairy cows' performance, with particular reference to fatty acids and fat-soluble antioxidants concentration. A three by three Latin square experiment was performed with 18 dairy cows. Experimental treatments consisted of exclusive feeding with grass silage and zero-grazing, both offered ad libitum indoors, or grazing for 24 h. The results showed that grazing cows had a higher dry matter intake and greater milk yield than cows feeding on grass silage and zero-grazing, as well as higher concentrations of protein, lactose, nonfat-solids and urea in milk than housed cows. Milk fat from grazing cows had a higher proportion of unsaturated fatty acids than from cows feeding on grass silage and zero-grazing, with significant differences in the proportion of vaccenic and rumenic acids. The 18:1 trans-11 to 18:1 trans-10 ratio is proposed as biomarker to identify the milk produced from the management system of grazing cattle. Milk from grazing cows had a greater proportion of lutein than cows eating grass silage, with the zero-grazing system having intermediate values. In conclusion, the mode of grass supply affects fatty acid and antioxidant profiles of milk.

Keywords: grass silage; zero-grazing; grazing; milk; antioxidants; fatty acids

1. Introduction

Milk has great importance for humans because of its nutritional characteristics, providing a high nutrient content in relation to its caloric content. The composition of milk determines its nutritional and industrial quality, which affects directly the profitability and competitiveness of milk production [1]. Nowadays, not only the nutritional value, but also other specific components of milk (e.g., fatty acids and vitamins) have attracted interest because of their important relevance in an overall aim to improve the long-term health of consumers and in the added value of milk and its products. Milk composition is the reflection of multiple factors that may or may not be modified through different livestock management practices. The concern of consumers about livestock foods and their methods of production related to the concept of food quality now include, in addition to nutritional value, flavor, aroma and color, as well as ethical aspects such as animal welfare and environment impact of the production system [2]. Consumers assume that a pasture-based dairy cattle feeding is more natural and better aligned with animal welfare requirements than more intensive systems. Knowing the production system can

---

determine the consumers choice at the time of purchase [3]. Thus, Monahan et al. [4] indicated the need to investigate biomarkers that allow associate milk composition with livestock food and, therefore, allow traceability from farm to fork according to production system and geographical origin.

Factors related to nutrition, feeding and management of dairy herds can produce changes in the short term [5]. Many studies show that milk produced from grazing systems has different characteristics in its components than milk produced from confinement systems [6,7]. Fresh forages are important natural source of antioxidants, vitamins and fatty acids in ruminant diets, and their concentrations in forage have an important relationship to the resultant composition and quality of milk and dairy products [8]. Liposoluble antioxidants and fatty acids in milk have been proposed as biomarkers for the authentication of milk produced from pastures, because they depend on external factors that differentiate the production system, diet, exercise and animal welfare [9].

The nutritional quality of milk fat is largely based on its fatty acid profile, which plays a key role in many vital functions and has a direct impact on the health of consumers [10]. In general, the fatty acid content of milk changes in quantity and quality depending on factors related mainly to diet and, to a lesser extent, to the animal and the environment [11–13], so that fat content and the fatty acid profile can be indicators of the diet of animals [14] and the management system [6].

The positive impact of liposoluble antioxidants (carotenoids and tocopherols) on human health has been extensively reviewed [15–18]. The concentration of these bioactive compounds in milk is directly related to dietary concentrations [19]. Fresh pasture is a good source of vitamins and antioxidants that are transferred to milk [20]. Higher levels of antioxidants (α-tocopherol, β-carotene and retinol) have been reported in milk from cows that consume fresh grass compared with diets rich in concentrate or silage [21,22].

The conditions of the oceanic climate in the Atlantic arc favor the production of grass and fodder to feed dairy cattle. Fresh or preserved forages are essential parts of the dairy cows' diets. A way to help the current profitability problem of milk producers could be guided by the feeding systems that enhance the utilization of their own forage resources. The models based on grazing allow cost savings concerning feeding on the farms [23,24]. Hanson et al. [25] found that dairy farms in the mid-Atlantic region of the USA based on extensive grazing were more profitable than housed dairy farms. However, not all animals have the possibility of grazing, either due to lack of available surface or due to handling difficulties. Thus many farmers have to adopt a cut-and-carry system or provide preserved grass to the manger [26,27].

The objective of the present study was to examine the effectiveness of the delivery method of herbage to dairy cows: grazing, cut-and-carry system (zero-grazing) or grass silage, on milk performance and milk antioxidant and fatty acids profiles with the view to identify biomarkers of the feeding system.

2. Materials and Methods

The work was undertaken at the Servicio Regional de Investigación y Desarrollo Agroalimentario (SERIDA) experimental farm (43°28′20″ N, 5°26′10″ W; 10 m above sea level) and adhered to the standard of the EU Animal Welfare Directive Number 2010/63/EU.

2.1. Experimental Design and Treatments

Eighteen Holstein cows in the second half of lactation were randomly distributed into three groups of six cows. All cows were selected with an initial average weight of 624 ± 69 kg (average ± standard error), 2.5 ± 1.5 lactations, 108 ± 53 days of lactation and an average daily production of 30.3 ± 7.1 kg of milk. Three treatments were tested: (1) ad libitum grass silage with housed cows (2) ad libitum zero-grazing indoors, and (3) full-time grazing. All treatments were supplemented daily with 6 kg of concentrate offered during milking to meet the energy requirements. Three trials were carried out by following a three by three Latin square design. Each trial lasted 19 days, including 13 days of adaptation to the experimental treatment and six days of sampling and measurements. In each trial, six animals received randomly ad libitum grass silage indoors plus 6 kg of concentrate; ad libitum fresh grass as zero-grazing plus 6 kg of concentrate; or grazing 24 h per day plus 6 kg of concentrate per animal per day with 18 replications per treatment. All forages offered indoor were supplied once a day, and concentrate was offered twice a day.

2.2. Experimental Procedure

Grazing and a cut-and-curry system were carried out in three plots with rotational management. The plot size was 1.5 ha with a wide range of grasses: *Poa trivialis* L. (23.15%), *Lolium perenne* L. (16.38%), *Holcus lanatus* L. (14.87%), *Dactylis glomerata* L. (14.26%), *Festuca arundinacea* Scherb. (2.48%), *Agrostis capillaries* L. (1.68%) and *Trisetum flavescens* P. Beauv. (1.21%); legumes: *Trifolium repens* L. (11.80%); species of the family Ranunculaceae: *Ranunculus bulbosus* L. (2.26%), *Ranunculus acris* L. (1.34%); species of the family Asteraceae: *Taraxacum officinale* Weber (1.56%), *Carlina* sp. (1.11%) and other species, all < 1.00% such as *Bromus hordeaceus* L., *Cerastium fontanum* Baumg., *Geranium mole* L., *Poa pratensis* L., *Galium verum* L., *Phleum pratense* L., *Bellis perennis* L., *Carex* sp., *Potentilla erecta* Raeusch., *Tragopogon pratensis* L., *Veronica chamaedrys* L., *Cerastium glomeratum* Thuill., *Lactuca* sp., *Dianthus monspessulanus* L. and *Sonchus soleraceus* L. that together reached 3.09%. Dead matter accounted for the remaining 4.81%. No grass was in bloom. The grass silage offered was from the herbage harvested during the previous spring in the same meadows.

During the experimental period, plots for grazing and zero-grazing were assigned to the corresponding group of animals taking into account the available pregrazing herbage mass. Herbage in the grazing treatment was sampled on the first day of the sampling period by tracing a diagonal transect across the paddock. The sample was composed of five squares (0.20 m^2 each), leaving a stubble of about 5 cm. Herbage of zero-grazing and grass silage treatments were sampled daily during the experimental period and pooled in one sample by period. Samples of concentrate were taken at the beginning of each experimental period.

2.3. Sampling and Chemical Analyses

The individual intake of grass in zero-grazing and silage treatments was automatically recorded daily by an electronic weighing system integrated with a scale pen by a computerized system. Herbage intakes on pasture were estimated using the animal performance method [28]. Briefly, energy requirements were recorded as net energy requirements for maintenance, lactation, body weight changes, walking and grazing. The net energy from herbage intake was estimated as total energy requirements minus the net energy supplied by concentrate intake. Concentrate intake was recorded daily by an automatic feeder coupled to the milking system.

All cows were milked twice daily (at 7:00 and 19:00 h). Milk production was measured and sampled daily in both milking sessions. The two samples from each cow were mixed proportionally according to the milk produced in both milkings.

Forage samples were dried at 60 °C for 24 h to determine the dry matter content, and ground at 0.75 mm. Concentrate samples were milled at 1.00 mm. Feed samples were analyzed for organic matter (OM), crude protein (CP), and neutral detergent fiber (NDF) by near infrared spectroscopy (FOSS NIRSystem 5000, Silver Springs, MD, USA). The energy content was estimated in all samples according to National Research Council (NRC) [29]. The fatty acid and antioxidants concentration of feedstuffs were analyzed in the Dairy Interprofessional Laboratory of Galicia (LIGAL). Extraction and methylation of the fatty acids (FA) were carried out simultaneously according to Sukhija and Palmquist methodology [30]. Esterification of FA was performed using a toluene and methanolic hydrochloric acid solution as follows: heating at 70 °C in a water bath for 2 h, cooling at room temperature and adding 2 mL of hexane and 5 mL of K_2CO_3 (6% *w/v*) and centrifuging for 5 min at 2500 rpm. The organic phase was immediately evaporated in a nitrogen stream to obtain an oily residue and dissolved in 0.8 mL of hexane. FA methyl esters were separated, identified and quantified using a TRACE GC Ultra equipment (Thermo Fisher Scientific, Waltham, MA, USA) with flame ionization detector (FID), using a

100 m × 0.25 mm i.d. fused silica capillary column (SP-2560 Capillary GC Column, Sigma-Aldrich Inc., Saint Louis, MO, USA). Helium was used as carrier gas at a flow rate of 0.6 mL/min. The temperature of the injector and detector were 250 and 260 °C, respectively. The injection volume was 1 µL. The initial column temperature was set at 140 °C for 5 min; from 140 to 200 °C at 3 °C/min and held for 5 min; from 200 to 240 °C at 3 °C/min and held for 5 min and, finally, held for 38 min. Individual FA were quantified through internal calibration using methylated 9:0, 17:1, 19:0 and 20:2 fatty acids as internal standards.

The samples for antioxidant assay were immediately vacuum packed and frozen (−20 °C) and analyzed according to Chauveau-Duriot et al. [31]. Samples were treated with liquid nitrogen in a Robot Coupe R6 grinder (Vincennes, France). The processing of samples was performed in dim light, using opaque glass. Butylhydroxytoluene (0.1% *v/v*) was added as antioxidant, $NaHCO_3$ as neutralizing agent, and 10 ppm of echinenone and 3 ppm of δ-tocopherol as internal standards. The lipophilic components were extracted by washing three times with acetone. The analytes were extracted with petroleum ether, the organic phase was evaporated under a nitrogen stream and the dry residue was saponified with KOH in MeOH (5.5% *w/v*) for 15 min at room temperature. After centrifugation at 1000× *g* for 5 min at room temperature, the organic phase was collected, evaporated again and reconstituted in the mobile phase. Finally, it was filtered through a syringe filter (Acrodisc Syringe Filter GHP, 25 mm, 0.2 µm, Waters, MA, USA) and transferred into a high performance liquid chromatography (HPLC) vial. An HPLC (Alliance 2695, Waters, MA, USA) system equipped with two serial detectors, UV-Vis and fluorescence, was used for the simultaneous detection and separation of xanthophylls, carotenes and vitamins A and E. Separation of antioxidant was carried out using a reverse phase column RP C18 Kinetex 2.6 µm 4.6 × 150 mm (Phenomenex, Torrance, CA, USA). The sample and column were kept refrigerated at 10 and 13 °C, respectively. The elution of components in the column was performed using a flow of 0.6 mL/min and a quaternary gradient of mobile phase. Quantification was carried out using external calibration models, quantifying the fat-soluble antioxidants according to the recovery factor of both internal standards.

Milk samples were analyzed for fat, protein, lactose, nonfat solids and urea by MilkoScan FT6000 (FOSS Tecctor, Millerd, Denmark) in the Dairy Interprofessional Laboratory of Asturias (LILA). The milk samples for fatty acid and antioxidants analysis were immediately frozen (−20 °C) until their analysis in the Dairy Interprofessional Laboratory of Galicia (LIGAL). Fatty acids were analyzed according to standard methods ISO14156:2001/IDF172 for lipids extraction and ISO15884:2002/IDF182 for preparation of fatty acid methyl esters. Twenty mL of milk were mixed with 96% ethanol, 30% ammonia solution and diethyl ether. After shaking for 1 min, the mixture was left to stand to achieve phase separation then hexane was added, mixed carefully, and left to stand for a second phase separation. Finally, the aqueous layer was discarded. Sodium sulfate solution (10% *w/v*) was added to the remaining content, mixed carefully again, left to stand for a third phase separation and, thereafter, the aqueous layer was discarded. The organic layer was transferred to a conical flask, mixed with anhydrous sodium sulfate, left to stand for 10 min and filtered. Finally, the filtrate was evaporated in a rotary steamer (Buchi R-114, Flawil, Switzerland) under a nitrogen stream in a water bath set at 50 °C. The extract was dissolved in hexane, saponified as described above and 0.5 g of sodium hydrogen sulfate were added. Finally, it was centrifuged at 1000× *g* for 5 min at room temperature. FA methyl esters were separated, identified and quantified using a Varian 3900 GC (Varian Inc., Palo Alto, CA, USA) with a flame ionization detector (FID), using a 120 m × 0.25 mm i.d. capillary column (BPX70 GC column, Thermo Fisher Scientific, Waltham, MA, USA). Helium was used as carrier gas at a flow rate of 1.3 mL/min. The temperature of the injector and detector were 250 °C. The initial column temperature was 45 °C for 5 min; from 45 to 175 °C at 13 °C/min and held for 27 min; from 175 to 215 °C at 4 °C/min and held for 35 min. Individual FA peaks were identified by comparison of their retention times with those of pure methyl ester standards (Supelco 37 Component FAME Mix and TVA methyl standard of Supelco Inc., Saint Louis, MO, USA, and methyl CLAc9t11 of Matreya LCC., State College, PA, USA). Individual FA were quantified using an internal calibration using methylated 9:0, 17:1c10, 18:2c12t10 (Matreya LCC., State College, PA, USA) and 19:0 (Sigma-Aldrich, Saint Louis, MO, USA).

Carotenoids and vitamins E and A present in the milk were determined according to the methodology proposed by Gentili et al. [32]. Milk samples were thawed the day before analysis and tempered previous to simultaneous extraction of carotenoids and vitamins. Sample preparation and identification and quantification of antioxidants were carried out according to the methodology described in foods antioxidants analysis.

2.4. Statistical Analysis

Statistical analysis was performed using the R statistical package [33]. Dry matter intake, milk yield and milk composition data were analyzed by a general linear model (GLM) procedure using repeated measures ANOVA by following a 3 × 3 Latin square design. The mode of supply of herbage (grazing, zero-grazing or silage grass) was considered as the fixed effect, and period and animal were considered as random effects. Individual animals were considered as experimental units. Significance was set at $p < 0.05$. When the ANOVA was significant, means were separated by Tukey's test pairwise comparison.

3. Results

Table 1 shows the nutritive composition of fresh forage used in grazing and zero-grazing treatments, and grass silage and concentrate used in all treatments. It was not possible to complete the fatty acid and antioxidant profiles analysis of grass used in zero-grazing treatment due to a problem with the samples' conservation, so these results were not included. The average of crude protein in fresh forage was higher than grass in zero-grazing and grass silage. The energy content of grass silage was lower than grass used in grazing and zero-grazing treatments. In general, the fatty acid and antioxidant concentrations were higher in fresh herbage than grass silage.

Table 1. Nutritive value, metabolizable energy, fatty acid and antioxidants profiles of forage used in grazing, zero-grazing, grass silage treatments and concentrate used in all treatments.

Item	Grazing	Zero-Grazing	Grass Silage	Concentrate
Dry matter (DM, %)	21.18	24.04	30.52	87.71
Organic matter (% DM)	87.45	91.15	89.47	91.43
Crude protein (% DM)	15.12	10.86	9.86	20.33
Neutral detergent fiber (% DM)	57.40	58.97	67.87	19.31
Metabolizable energy (MJ/kg DM)	9.56	9.29	8.27	12.74
Fatty Acids (g/100 g fatty acids)				
10:1 cis-9	0.28	NA [1]	0.06	0.02
11:0	0.21	NA	0.84	0.01
12:0	0.70	NA	1.12	0.18
13:0	0.86	NA	0.88	0.01
14:0	0.62	NA	0.80	0.49
15:0	0.14	NA	0.33	0.05
15:1 cis-10	1.33	NA	1.14	0.00
16:0	18.19	NA	19.97	25.33
16:1 cis-7 + 16:1 trans-9	2.31	NA	1.43	0.05
16:1 cis-9	0.27	NA	1.16	0.14
17:0	0.28	NA	0.51	0.10
18:0	1.98	NA	2.61	2.77
18:1 cis-9	3.74	NA	8.98	28.09
18:2 cis-9 cis-12	16.50	NA	28.22	39.15
18:3 cis-6 cis-9 cis-12	0.11	NA	0.19	0.05
18:3 cis-9 cis-12 cis-15	49.05	NA	26.76	2.37
20:0	0.73	NA	0.87	0.29

Table 1. *Cont.*

Item	Grazing	Zero-Grazing	Grass Silage	Concentrate
20:1 cis-9	0.24	NA	0.32	0.25
20:1 cis-11	0.31	NA	0.50	0.35
21:0	0.17	NA	0.32	0.05
22:0	1.08	NA	1.49	0.13
23:0	0.15	NA	0.35	0.03
24:0	0.65	NA	1.05	0.09
24:1	0.11	NA	0.10	0.01
Σ SFA [2]	25.75	NA	31.14	29.52
Σ MUFA [3]	8.58	NA	13.69	28.91
Σ PUFA [4]	65.67	NA	55.17	41.56
PUFA to SFA ratio	2.55	NA	1.77	1.41
n-6 to n-3 ratio	0.34	NA	1.05	16.54
Antioxidants (mg/kg DM)				
Neoxanthin	14.97	NA	0.44	0.04
Violaxanthin	13.28	NA	0.29	<LQ [5]
Antheraxanthin	1.58	NA	0.95	0.01
Lutein	62.45	NA	23.79	0.43
Zeaxanthin	3.21	NA	1.49	0.07
Β-Cryptoxanthin	0.57	NA	0.18	0.04
Σ-trans-β-Carotenes	30.81	NA	7.17	0.09
9-cis-β-Carotenes	6.28	NA	2.13	0.06
13-cis-β-Carotenes	3.44	NA	0.82	0.08
α-tocopherol	9.64	NA	8.45	2.83
γ-tocopherol	1.60	NA	1.03	4.29

[1] not analyzed. [2] Saturated fatty acids. [3] Monounsaturated fatty acids. [4] Polyunsaturated fatty acids. [5] <LQ, below quantification level.

The results of dry matter intake and milk production and composition are presented in Table 2. The dry matter intake of forage was higher ($p < 0.001$) in the grazing system than in both housed systems. There were no differences in dry matter intake of concentrate among treatments. The yield was different among systems ($p < 0.001$). The greatest milk production was observed in the grazing feeding system, with intermediate values in the zero-grazing, and the lowest in the grass silage treatment. Grazing milk had the highest proportion of protein, lactose, nonfat solids and urea ($p < 0.001$) compared to both treatments with housed cows. There were no differences in fat proportion among feeding systems.

Table 2. Food intake, milk yield and milk composition according to mode of supply of herbage to dairy cows: grazing, zero-grazing or grass silage.

Item	Grazing	Zero-Grazing	Grass Silage	RSD	p [2]
Forage (kg DM/d)	14.34 [a]	11.54 [b]	10.54 [b]	2.40	***
Concentrate (kg DM/d)	3.75	3.66	3.56	0.62	NS
Milk (kg/d)	23.4 [a]	18.1 [b]	14.0 [c]	3,57	***
Fat (g/kg)	35.8	33.7	36.1	3.18	NS
Protein (g/kg)	32.1 [a]	29.1 [b]	27.8 [b]	2.90	***
Lactose (g/kg)	45.7 [a]	41.0 [b]	41.7 [b]	2.56	***
SNF [1] (g/kg)	83.9 [a]	77.8 [b]	76.3 [b]	3.79	***
Urea (mg/kg)	281 [a]	200 [b]	215 [b]	40.3	***

[1] Solids nonfat. [2] Statistical significance: NS: $p > 0.05$, *: $p < 0.05$, **: $p < 0.01$, ***: $p < 0.001$. [a,b,c] Means followed by different lowercase letters are significantly different.

The milk fatty acid profile is shown in Table 3. Grazing cows presented a higher proportion of vaccenic acid (18:1 trans-11, $p < 0.05$) and rumenic acid (18:2 cis-9 trans-11, $p < 0.01$), as well as a higher ($p < 0.05$) 18:1 trans-11 to 18:1 trans-10 ratio than zero-grazing and grass silage treatments. The milk of the silage system had a higher ($p < 0.05$) proportion of linoleic acid (18:2 cis-9 cis-12) than the grazing system, while there were no differences among treatment in α-linolenic acid (18:3 cis-9 cis-12 cis-15) proportions. There were no differences in the proportion of monounsaturated, saturated and polyunsaturated fatty acids, between unsaturated and saturated fatty acids or n-6 to n-3 ratios among treatments.

Table 3. Fatty acids profile in milk according to mode of supply of herbage to dairy cows: grazing, zero-grazing or grass silage.

Item (g/100 g Fatty Acids)	Grazing	Zero-Grazing	Grass Silage	SD [1]	p [2]
4:0	5.57	5.36	5.31	0.286	NS
6:0	1.84 [a]	1.89 [a]	1.71 [b]	0.053	*
8:0	0.97	0.99	0.85	0.061	NS
10:0	2.07	2.11	1.71	0.223	NS
10:1 cis-9	0.06	0.06	0.05	0.005	NS
11:0	0.03	0.02	0.01	0.009	NS
12:0	2.14	2.45	1.97	0.271	NS
13:0	0.09	0.09	0.08	0.009	NS
14:0	9.82	10.14	8.98	0.699	NS
14:0 iso	0.22	0.20	0.21	0.010	NS
14:1 cis-9	1.00	1.13	0.92	0.104	NS
15:0	1.21	1.24	1.17	0.046	NS
15:0 iso	0.46	0.47	0.42	0.035	NS
15:0 anteiso	0.74 [a]	0.71 [a]	0.59 [b]	0.040	*
15:1 cis-10	0.01	0.01	0.01	0.003	NS
16:0	27.86	30.73	29.94	1.031	NS
16:1 cis-9	1.80	2.17	2.20	0.180	NS
17:0	0.58	0.64	0.65	0.028	NS
18:0	9.89	9.57	8.85	0.746	NS
18:1 trans-6 + 18:1 trans-9	0.50 [a]	0.41 [b]	0.39 [b]	0.024	*
18:1 trans-10	0.27 [a]	0.21 [b]	0.20 [b]	0.025	*
18:1 trans-11	5.08 [a]	2.77 [b]	2.01 [b]	0.745	*
18:1 trans-12	0.20 [a]	0.15 [b]	0.14 [b]	0.015	*
18:1 cis-9	21.59	21.97	25.52	1.521	NS
18:1 cis-11	0.37 [b]	0.44 [b]	0.57 [a]	0.051	*
18:1 cis-12	0.06 [b]	0.06a [b]	0.07 [a]	0.004	*
18:2 tran-9 trans-12	0.07 [a]	0.07 [a]	0.05 [b]	0.003	**
18:2 cis-9, cis-12	1.16 [b]	1.38 [ab]	1.60 [a]	0.124	*
18:2 cis-9 trans-11 (CLA [3])	2.29 [a]	1.37 [b]	1.05 [b]	0.218	**
Other isomers CLA	0.21	0.24	0.24	0.015	NS
18:3 cis-9 cis-12 cis-15	0.55	0.47	0.43	0.070	NS
18:3 cis-6 cis-9 cis-12	0.03	0.03	0.03	0.003	NS
20:0	0.21 [b]	0.25 [a]	0.27 [a]	0.012	**
20:3	0.15 [b]	0.22 [ab]	0.27 [a]	0.030	*
20:5	0.01	0.01	0.01	0.006	NS
20:2	0.01 [c]	0.02 [b]	0.03 [a]	0.001	***
20:3	0.08	0.10	0.10	0.011	NS
20:4	0.01	0.02	0.02	0.004	NS
20:1 cis-11	0.04	0.03	0.05	0.006	NS
21:0	0.07	0.08	0.08	0.005	NS
22:0	0.07	0.08	0.07	0.006	NS
22:5	0.10 [b]	0.11 [b]	0.15 [a]	0.012	*
22:6	0.01	0.01	0.01	0.009	NS
22:2	0.08	0.09	0.10	0.013	NS
23:0	0.05	0.06	0.05	0.006	NS
24:0	0.07	0.08	0.07	0.006	NS
Sum of fatty acids					
$\sum$ SFA [4]	62.81	65.07	62.49	1.479	NS
$\sum$ BCFA [5]	1.43 [a]	1.39 [a]	1.22 [b]	0.053	*
$\sum$ MUFA [6]	30.98	29.43	32.16	1.417	NS
$\sum$ cis-MUFA	24.93	25.89	29.41	1.743	NS
$\sum$ trans-MUFA	6.05 [a]	3.53 [b]	2.75 [b]	0.791	*
$\sum$ PUFA [7]	4.78	4.11	4.13	0.271	NS
$\sum$ n-6	1.45 [b]	1.70 [ab]	1.92 [a]	0.142	*
$\sum$ n-3	0.83	0.79	1.00	0.062	NS
Ratios					
PUFA to SFA	0.08	0.06	0.07	0.007	NS
UFA to SFA	0.57	0.52	0.58	0.037	NS
18:1 trans-11 to 18:1 trans-10	18.34 [a]	13.08 [b]	10.35 [b]	1.912	*
n-6 to n-3	1.77	2.17	2.10	0.189	NS

[1] Standard deviation. [2] Statistical significance: NS: $p > 0.05$, *: $p < 0.05$, **: $p < 0.01$, ***: $p < 0.001$. [3] Conjugated linoleic acid. [4] Saturated fatty acids. [5] Branched chain fatty acids [6] Monounsaturated fatty acids. [7] Polyunsaturated fatty acids. [a], [b], [c] Means followed by different lowercase letters are significantly different.

The content of fat-soluble antioxidants according to the mode of herbage supply to cows is shown in Table 4. No effect of feeding method was observed in vitamins A and E in milk. Milk from grazing cows had a greater proportion of lutein than milk from cows offered grass silage ($p < 0.01$), with the zero-grazing system showing intermediate values. There were no differences among treatment in other carotenoids and carotenes.

Table 4. Fat-soluble antioxidants composition according to mode of supply of herbage to dairy cows: grazing, zero-grazing or grass silage.

Item (μg/L Milk)	Grazing	Zero-Grazing	Grass Silage	SD [1]	p [2]
Retinol	855	852	827	202.6	NS
α-Tocopherol	1189	962	1068	237.5	NS
γ-Tocopherol	17.8	19.9	23.3	3.50	NS
Lutein	21.9 [a]	15.5 [b]	9.1 [c]	3.19	**
Zeaxanthin	1.19	0.42	0.47	0.332	NS
β-Cryptoxanthin	3.13	1.64	1.55	1.031	NS
All-trans-β-Carotene	255	184	179	45.6	NS
9-cis-β-Carotene	1.73	2.05	1.60	0.775	NS
13-cis-β-Carotene	9.52	7.08	6.82	1.540	NS

[1] Standard deviation. [2] Statistical significance: NS: $p > 0.05$, *: $p < 0.05$, **: $p < 0.01$, ***: $p < 0.001$. [a],[b],[c] Means followed by different lowercase letters are significantly different.

4. Discussion

The model of milk production has changed in the western world in the last years and involves intensification of production with an increase in inputs that is not reflected in an increase in the price and quality of milk [34]. Although prices and availability of feed ingredients vary by regions, in the Atlantic arc the production of pastures and legumes used to feed dairy cattle is a great way to decrease the feed diets' expenses [35].

Energy intake is often a primary limitation for milk yield in grazing cows, even for pastures with high-quality energy content [36]. A good diet formulation strategy should be determined according to ingredient availability. In this study, herbage was used as the only forage source in order to avoid other variation factors. Forage is rarely offered as the sole feed to lactating dairy cows because the dry matter intake (DMI) is generally too low to meet the requirements. The theoretical DMI capacity of dairy cows is approximately 21 kg DM/day [29]. According to our own previous experience, around 15 kg DM/day is the voluntary intake of forage for this type of animal. Consequently, six kilograms (fresh matter basis) of concentrate were offered daily in order to meet the energy requirements for dairy cows with these features (approximately 210 MJ metabolizable energy per day [29]) and a forage to concentrate ratio of 75:25. Nevertheless, only the cows in the grazing treatment reached a forage intake up to 14 kg DM/day, while indoor treatments barely reached 11 kg DM/day. The lower DMI of grass silage treatment than that of the grazing treatment has been previously reported [37], because voluntary intake of silage is lower than that of the herbage from which it was produced. The higher DMI in grazing cows compared with zero-grazing cows could be associated with a greater chance of grazing cows to select better botanical species and the best parts of plants, while housed zero-grazing cows do not have that opportunity. Grazing cows can more readily reject unsuitable forage and preferentially consume more easily digestible portions of grass. Furthermore, the dairy cows consumed just the half of the concentrate offered in all of the treatments tested. The concentrate was offered to the animals twice daily (7:00 and 19:00 h) at the time of milking. This fact could have induced a ruminal subacute acidosis that could lead to a drop in the intake of forage subsequently offered, in spite of maintaining the forage to concentrate ratio. With these levels of intake, is not possible to meet the energetic requirements for cows producing 30 kg milk/day. The energy deficit has been quantified as 25, 53 and 75 MJ/d for grazing, zero-grazing and grass silage treatments, respectively. For this reason, there was a decrease in milk production observed with respect to the values before the beginning of the assay.

The grass supply modes studied in this experiment exhibited a great variation in milk production, milk composition and fatty acid and antioxidant profiles, particularly in lutein content, which are strongly related to the specific compounds from feeds. Grazing cows had a higher DMI than housed cows offered zero-grazing or grass silage. These differences had a significant effect on milk production and milk composition, with a higher concentration of protein, lactose and nonfat solids in the milk of the grazing cows than in the housed cows. In contrast, other studies have shown lower DMI at pasture [38,39] and a consequent drop in milk yield in grazing cows compared to housed cows [40]. However, it should be taken into account that the housed cows in those studies were feeding with balanced total mixed rations while in our study grass or grass silage was the only ingredient of the ration. Furthermore, in our study conditions, grazing cows had the possibility to readily reject unsuitable forage and preferentially consume the leafy and more easily digestible portions of the grass with, possibly, better nutritional value [37]. Other authors showed that it could be possible to maximize energy intake including forages with high NDF digestibility which minimize filling effects and increase milk yield [35,41]. The protein concentration of milk in grazing cows was unchanged with respect to the initial situation, presumably because energy intake was similar to energy requirements, and may be due to a higher duodenal flow of microbial protein and total amino acids [37]. However, grazing cows have a greater concentration of urea in milk than in indoor treatments, in all of the cases in the normal range. This can be explained because the high concentration of protein in meadow grass could have increased the rate of microbial protein synthesis, as well as the concentration of propionate in the rumen, resulting in an increase in milk protein [42]. This can be explained because grazing animals would first select green shoots and the best parts of the plant, which can have great rumen degradability of protein [36], so an excess of urea could accumulate in the rumen that would be absorbed and excreted.

It is well-known that the concentration of lactose is rarely influenced by feeding. However, the results showed a higher lactose concentration in milk from grazing cows. A possible explanation might be related to the increased forage proportion, of up to 80%, which may lead to proliferation of cellulolytic rumen bacteria leading to more propionic acid and, eventually, an increase in the concentration of milk lactose content [43]. Increasing propionate is essential for promoting energy availability for milk production and increasing glucose and lactose synthesis [44]. In addition, as fresh herbage contains a high concentration of sugar, the synthesis of lactic acid in the rumen is favored, which in turn results in a high lactose content in milk. According to the study of Argamentería et al. [45], an increase in the energy supplied in the diet can be associated with an increase in the proportion of lactose.

Some authors have observed that milk produced by grazing cows has a higher fat content than milk produced under a semi-extensive or intensive system [46,47], although other studies have shown opposite results [5,36]. Among hundreds of fatty acids present in bovine milk fat, just a limited number is important because of their relation to nutritional, sensorial and technological properties [48]. The proportion of some transunsaturated fatty acids in bovine milk fat ranges from 2% to 8% of total fatty acids. Among them, the most interest has been focused mainly on transvaccenic acid and conjugated linoleic acid (CLA), and especially rumenic acid because it is the most abundant CLA isomer, for their important antisclerotic, antiatherogenic and anticarcinogenic properties [48,49].

Variations among treatments in the milk fatty acid profile were found. A higher proportion of transvaccenic and rumenic acids, as well as a higher 18:1 trans-11 to 18:1 trans-10 ratio, could be explained in grazing treatments by the proportion of fresh forage in the diet, which is associated with a higher intake of polyunsaturated fatty acids (PUFA) from fresh grass [50]. In addition, an increase in the losses of unsaturated fatty acids (UFA) and total fatty acids (FA) is produced during the wilting and silage processes [48]. Furthermore, the biohydrogenation of 18:2 n-6 could be affected by the high PUFA content in the rumen [51] and, as a consequence, could increase with higher contents of these fatty acids [14]. These results show that the fatty acid profile varies depending on the production system, with a greater proportion of UFA in cows fed on pasture [5]. A direct relationship in the proportion of UFA and decrease in the proportion of saturated fatty acids (SFA) is associated with increasing the proportion of fresh pasture in the cow diet [20]. Differentiation between milk from cows fed fresh grass indoors (zero-grazing) and grazing outdoors has proven difficult. Milk FA profile is different and may be healthier in grazing cows than when cows are stall-fed [52]. Lesser proportions of CLA and 18:1 trans FA in milk from zero-grazing compared to grazing treatments could be explained by the lower DMI. In addition, PUFA losses have been reported in grass immediately after cutting in fresh and preserved forages by oxidative processes of plant tissues [48]. Therefore, fatty acid intake is affected. Other factors such as intake pattern, or possibilities for feed selection by the cow, could also play a role. In our study higher trans isomers of monounsaturated fatty acids (MUFA) proportion and higher 18:1 trans-11 to 18:1 trans-10 ratio have been observed in milk from grazing cows compared to zero-grazing treatments. The 18:1 trans-11 to 18:1 trans-10 ratio increases when fresh forage proportion increases in the diet [53]. Consequently, the 18:1 trans-11 to 18:1 trans-10 ratio can be proposed to identify grass-fed milk.

Fat-soluble antioxidants and vitamins present in cows' milk are derived specially from green forage [54,55]. Milk and dairy products are a rich source of carotenoids and bioavailable vitamin A in the daily diet of consumers [56]. However, they have low levels of vitamin E. Vitamin E plays an important role as an antioxidant that protects milk fat against autoxidation [57]. Beneficial effects related to the reduction of oxidative stress, which was shown to be a risk factor for a wide range of chronic disease processes including cardiovascular disease, cancer, neurodegenerative diseases, impaired immunity and premature ageing, are associated with consumption of diets with high antioxidant contents [17,56].

The variability in milk content of carotenoids and vitamins has been associated with the presence of grass in the ration and its levels in the grass are highly related to drying and preservation because of light exposure [22]. Different studies reported that the content of β-carotene and fat-soluble vitamins were up to four times higher in the milk of grazing cows compared to the milk of cows offered total mixed rations or a high proportion of concentrate [56,58]. Milk produced from pasture is yellower in color as a result of the higher β-carotene concentration of the milk [58]. Our results did not show wide differences in the antioxidant profile among treatments. Lutein was the only antioxidant in milk that showed significant differences. It could be because all treatments were based on meadows with very similar botanical species and a low biodiversity.

5. Conclusions

In conclusion, the modification of milk composition is associated with the feeding system. It is possible to distinguish from cows in a grazing feeding system by significant variability in fatty acids profile, as well as lutein content. Both were associated with the presence of fresh grass in the diet, especially when grass is consumed as grazing. Similarly, under the conditions of the experiment, grazing dairy cows had higher milk production and higher concentrations of protein, lactose, nonfat solids, urea and lutein, as well as unsaturated fatty acids. The 18:1 trans-11:18 to 1 trans-10 ratio is proposed as biomarker to identify milk produce from the management system of grazing cattle.

Author Contributions: Conceptualization, F.V.; methodology, F.V., A.M.-F. and C.C.; software, S.D.L.T.-S. and F.V.; validation, F.V., A.M.-F. and L.J.R.; formal analysis, S.D.L.T.-S., A.M.-F. and C.C.; investigation, S.D.L.T.-S. and F.V.; resources, F.V., A.M.-F. and L.J.R.; data curation, S.D.L.T.-S. and F.V.; writing—original draft preparation, S.D.L.T.-S.; writing—review and editing, F.V., A.M.-F. and L.J.R.; visualization, S.D.L.T.-S. and F.V.; supervision, F.V., A.M.-F. and L.J.R.; project administration, L.J.R.; funding acquisition, L.J.R. All authors have read and agreed to the published version of the manuscript.

Funding: Work financed by National Institute for Agricultural and Food Research and Technology (INIA, Spain) through project RTA2014-00086-C02, by the Principality of Asturias through PCTI IDI2018-000237 (GRUPIN NySA) and co-financed with European Regional Development Fund (ERDF). S. De La Torre-Santos is recipient of a SENACYT-IFARHU, Panama, doctoral fellowship.

Acknowledgments: We would like to offer our special thanks to SERIDA experimental farm and animal nutrition laboratory staffs.

Conflicts of Interest: The authors declare no conflict of interest.

References

1. Chilliard, Y.; Glasser, F.; Enjalbert, F.; Ferlay, A.; Bernard, L.F.; Rouel, J.; Doreau, M. Diet, rumen biohydrogenation and nutritional quality of cow and goat milk fat. *Eur. J. Lipid Sci. Technol.* **2007**, *109*, 828–855. [CrossRef]
2. Luykx, D.M.; Van Ruth, S.M. An overview of analytical methods for determining the geographical origin of food products. *Food Chem.* **2008**, *107*, 897–911. [CrossRef]
3. Valenti, B.; Martin, B.; Andueza, D.; Leroux, C.; Labonne, C.; Lahalle, F.; Larroque, H.; Brunschwig, P.; Lecomte, C.; Brochard, M.; et al. Infrared spectroscopic methods for the discrimination of cows' milk according to the feeding system, cow breed and altitude of the dairy farm. *Int. Dairy J.* **2013**, *32*, 26–32. [CrossRef]
4. Monahan, F.J.; Moloney, A.P.; Downey, G.; Dunne, P.G.; Schmidt, O.; Harrison, S.M. Authenticity and traceability of grassland production and products. *Grassl. Sci. Eur.* **2010**, 401–414.
5. Morales-Almaráz, E.; Soldado, A.; González, A.; Martínez-Fernández, A.; Domínguez-Vara, I.A.; de la Roza-Delgado, B.; Vicente, F. Improving the fatty acid profile of dairy cow milk by combining grazing with feeding of total mixed ration. *J. Dairy Res.* **2010**, *77*, 225–230. [CrossRef]
6. Morales-Almaráz, E.; de la Roza-Delgado, B.; González, A.; Soldado, A.; Rodríguez, M.L.; Peláez, M.; Vicente, F. Effect of feeding system on unsaturated fatty acid level in milk of dairy cows. *Renew. Agric. Food Syst.* **2011**, *26*, 224–229. [CrossRef]
7. Kusche, D.; Kuhnt, K.; Ruebesam, K.; Rohrer, C.; Nierop, A.F.; Jahreis, G.; Baars, T. Fatty acid profiles and antioxidants of organic and conventional milk from low-and high-input systems during outdoor period. *J. Sci. Food Agric.* **2015**, *95*, 529–539. [CrossRef] [PubMed]
8. Elgersma, A.; Søegaard, K.; Jensen, S.K. Fatty acids, α-tocopherol, β-carotene, and lutein contents in forage legumes, forbs, and a grass–clover mixture. *J. Agric. Food Chem.* **2013**, *61*, 11913–11920. [CrossRef] [PubMed]
9. Moloney, A.P.; Monahan, F.J.; Schmidt, O. Quality and authenticity of grassland products. *Grassl. Sci. Eur.* **2014**, *19*, 509–520.
10. Belury, M.A. Dietary conjugated linoleic acid in health: Physiological effects and mechanisms of action. *Annu. Rev. Nutr.* **2002**, *22*, 505–531. [CrossRef]
11. Hernández-Ortega, M.; Martínez-Fernández, A.; Soldado, A.; González, A.; Arriaga-Jordán, C.M.; Argamentería, A.; de la Roza-Delgado, B.; Vicente, F. Effect of total mixed ration composition and daily grazing pattern on milk production, composition and fatty acids profile of dairy cows. *J. Dairy Res.* **2014**, *81*, 471–478. [CrossRef]
12. Schwendel, B.H.; Wester, T.J.; Morel, P.C.H.; Tavendale, M.H.; Deadman, C.; Shadbolt, N.M.; Otter, D.E. Organic and conventionally produced milk-An evaluation of factors influencing milk composition. *J. Dairy Sci.* **2015**, *98*, 721–746. [CrossRef] [PubMed]
13. Morales-Almaráz, E.; de la Roza-Delgado, B.; Soldado, A.; Martínez-Fernández, A.; González, A.; Domínguez-Vara, I.A.; Vicente, F. Parity and grazing-time effects on milk fatty acid profile in dairy cows. *Anim. Prod. Sci.* **2018**, *58*, 1233–1238. [CrossRef]
14. Vicente, F.; Santiago, C.; Jiménez-Calderón, J.D.; Martínez-Fernández, A. Capacity of milk composition to identify the feeding system used to feed dairy cows. *J. Dairy Res.* **2017**, *84*, 254–263. [CrossRef] [PubMed]
15. Bendich, A. Physiological role of antioxidants in the immune system. *J. Dairy Sci.* **1993**, *76*, 2789–2794. [CrossRef]
16. Schneider, C. Chemistry and biology of vitamin E. *Mol. Nut. Food Res.* **2005**, *49*, 7–30. [CrossRef]
17. Willcox, J.K.; Ash, S.L.; Catignani, G.L. Antioxidants and prevention of chronic disease. *Crit. Rev. Food Sci. Nutr.* **2004**, *44*, 275–295. [CrossRef]
18. Tijerina-Sáenz, A.; Innis, S.M.; Kitts, D.D. Antioxidant capacity of human milk and its association with vitamins A and E and fatty acid composition. *Acta Paediatr.* **2009**, *98*, 1793–1798. [CrossRef]
19. Martin, B.; Fedele, V.; Ferlay, A.; Grolier, P.; Rock, E.; Gruffat, D.; Chilliard, Y. Effects of grass-based diets on the content of micronutrients and fatty acids in bovine and caprine dairy products. *Grassl. Sci. Eur.* **2004**, *9*, 876–886.
20. Alothman, M.; Hogan, S.A.; Hennessy, D.; Dillon, P.; Kilcawley, K.N.; O'Donovan, M.; Tobin, J.; Fenelon, M.A.; O'Callaghan, T.F. The "Grass-Fed" milk story: Understanding the impact of pasture feeding on the composition and quality of bovine milk. *Foods* **2019**, *8*, 350. [CrossRef]

21. Havemose, M.S.; Weisbjerg, M.R.; Bredie, W.L.P.; Nielsen, J.H. Influence of feeding different types of roughage on the oxidative stability of milk. *Int. Dairy J.* **2004**, *14*, 563–570. [CrossRef]

22. Agabriel, C.; Cornu, A.; Journal, C.; Sibra, C.; Grolier, P.; Martin, B. Tanker milk variability according to farm feeding practices: Vitamins A and E, carotenoids, color, and terpenoids. *J. Dairy Sci.* **2007**, *90*, 4774–4896. [CrossRef] [PubMed]

23. Soder, K.J.; Rotz, C.A. Economic and environmental impact of four levels of concentrate supplementation in grazing dairy herds. *J. Dairy Sci.* **2001**, *84*, 2560–2572. [CrossRef]

24. Peyraud, J.L.; Delagarde, R. Managing variations in dairy cow nutrient supply under grazing. *Animals* **2013**, *7*, 57–67. [CrossRef] [PubMed]

25. Hanson, J.C.; Johnson, D.M.; Lichtenberg, E.; Minegishi, K. Competitiveness of management-intensive grazing dairies in the mid-Atlantic region from 1995 to 2009. *J. Dairy Sci.* **2013**, *96*, 1894–1904. [CrossRef] [PubMed]

26. Martínez-García, C.G.; Dorward, P.; Rehman, T. Factors influencing adoption of improved grassland management by small-scale dairy farmers in central Mexico and the implications for future research on smallholder adoption in developing countries. *Livest. Sci.* **2013**, *152*, 228–238. [CrossRef]

27. Velarde-Guillén, J.; Estrada-Flores, J.G.; Rayas-Amor, A.A.; Vicente, F.; Martínez-Fernández, A.; Heredia-Nava, D.; Celis-Alvarez, M.D.; Aguirre-Ugarte, I.K.; Galindo-González, E.; Arriaga-Jordán, C.M. Supplementation of dairy cows with commercial concentrate or ground maize grain under cut-and-carry or grazing of cultivated pastures in small-scale systems in the highlands of central Mexico. *Anim. Prod. Sci.* **2019**, *59*, 368–375. [CrossRef]

28. Macoon, B.; Sollenberger, L.E.; Moore, J.E.; Staples, C.R.; Fike, J.H.; Portier, K.M. Comparison of three techniques for estimating the forage intake of lactating dairy cows on pasture. *J. Anim. Sci.* **2003**, *81*, 2357–2366. [CrossRef]

29. NRC. *Nutrient Requirements of Dairy Cattle*, 7th ed.; National Academic Press: Washington, DC, USA, 2001.

30. Sukhija, P.S.; Palmquist, D.L. Rapid method for determination of total fatty acid content and composition of feedstuffs and feces. *J. Agric. Food Chem.* **1988**, *36*, 1202–1206. [CrossRef]

31. Chauveau-Duriot, B.; Doreau, M.; Noziere, P.; Graulet, B. Simultaneous quantification of carotenoids, retinol, and tocopherols in forages, bovine plasma, and milk: Validation of a novel UPLC method. *Anal. Bioanal. Chem.* **2010**, *397*, 777–790. [CrossRef]

32. Gentili, A.; Caretti, F.; Bellante, S.; Ventura, S.; Canepari, S.; Curini, R. Comprehensive profiling of carotenoids and fat-soluble vitamins in milk from different animal species by LC-DAD-MS/MS hyphenation. *J. Agric. Food Chem.* **2013**, *61*, 1628–1639. [CrossRef] [PubMed]

33. R Core Team. *R: A Language and Environment for Statistical Computing. R Foundation for Statistical Computing*; R Core Team: Vienna, Austria, 2018; Available online: https://www.R-project.org/ (accessed on 16 July 2020).

34. Álvarez, A.; del Corral, J.; Solís, D.; Pérez, J.A. Does intensification improve the economic efficiency of dairy farms? *J. Dairy Sci.* **2008**, *91*, 3693–3698. [CrossRef] [PubMed]

35. Allen, M.S. Effects of diet on short-term regulation of feed intake by lactating dairy cattle. *J. Dairy Sci.* **2000**, *83*, 1598–1624. [CrossRef]

36. Bargo, F.; Muller, L.D.; Delahoy, J.E.; Cassidy, T.W. Performance of high producing dairy cows with three different feeding systems combining pasture and total mixed rations. *J. Dairy Sci.* **2002**, *85*, 2948–2963. [CrossRef]

37. Mohammed, R.; Stanton, C.S.; Kennelly, J.J.; Kramer, J.K.G.; Mee, J.F.; Glimm, D.R.; O'Donovan, M.; Murphy, J.J. Grazing cows are more efficient than zero-grazed and grass silage-fed cows in milk rumenic acid production. *J. Dairy Sci.* **2009**, *92*, 3874–3893. [CrossRef]

38. Bargo, F.; Muller, L.D.; Kolver, E.S.; Delahoy, J.E. Production and digestion of supplemented dairy cows on pasture. *J. Dairy Sci.* **2003**, *86*, 1–42. [CrossRef]

39. Lahlou, M.N.; Kanneganti, R.; Massingill, L.J.; Broderick, G.A.; Park, Y.; Pariza, M.W.; Fergunson, J.D.; Wu, Z. Grazing increases the concentration of CLA in dairy cow milk. *Animals* **2014**, *8*, 1191–1200. [CrossRef]

40. Vahmani, P.; Glover, K.E.; Fredeen, A.H. Effects of pasture versus confinement and marine oil supplementation on the expression of genes involved in lipid metabolism in mammary, liver, and adipose tissues of lactating dairy cows. *J. Dairy Sci.* **2014**, *97*, 4174–4183. [CrossRef]

41. Oba, M.; Allen, M.S. Effects of brown midrib 3 mutation in corn silage on productivity of dairy cows fed two levels of dietary NDF: 1. Feeding behavior and nutrient utilization. *J. Dairy Sci.* **2000**, *83*, 1333–1341. [CrossRef]

42. Couvreur, S.; Hurtaud, C.; Lopez, C.; Delaby, L.; Peyraud, J.L. The Linear Relationship between the Proportion of Fresh Grass in the Cow Diet, Milk Fatty Acid Composition, and Butter Properties. *J. Dairy Sci.* **2006**, *89*, 1956–1969. [CrossRef]

43. Lemosquet, S.; Delamaire, E.; Lapierre, H.; Blum, J.W.; Peyraud, J.L. Effects of glucose, propionic acid, and nonessential amino acids on glucose metabolism and milk yield in Holstein dairy cows. *J. Dairy Sci.* **2009**, *92*, 3244–3257. [CrossRef]

44. El-Zaiat, H.M.; Kholif, A.E.; Mohamed, D.A.; Matloup, O.H.; Anele, U.Y.; Sallam, S.M. Enhancing lactational performance of Holstein dairy cows under commercial production: Malic acid as an option. *J. Sci. Food Agric.* **2019**, *99*, 885–892. [CrossRef] [PubMed]

45. Argamentería, A.; Vicente, F.; Martínez-Fernández, A.; Cueto, M.A.; de la Roza-Delgado, B. Influence of partial total mixed rations amount on the grass voluntary intake by dairy cows. *Grassl. Sci. Eur.* **2006**, *11*, 161–163.

46. Capuano, E.; Boerrigter-Eenling, R.; Elgersma, A.; Van Ruth, S.M. Effect of fresh grass feeding, pasture grazing and organic/biodynamic farming on bovine milk triglyceride profile and implications for authentication. *Eur. Food Res. Technol.* **2014**, *238*, 573–580. [CrossRef]

47. Frétin, M.; Ferlay, A.; Verdier-Metz, I.; Fournier, F.; Montel, M.C.; Farruggia, A.; Delbes, C.; Martin, B. The effects of low-input grazing systems and milk pasteurisation on the chemical composition, microbial communities, and sensory properties of uncooked pressed cheeses. *Int. Dairy J.* **2017**, *64*, 56–67. [CrossRef]

48. Kalač, P.; Samková, E. The effects of feeding various forages on fatty acid composition of bovine milk fat: A review. *Czech J. Anim. Sci.* **2010**, *55*, 521–537. [CrossRef]

49. Parodi, P.W. Milk fat in human nutrition. *Aust. J. Dairy Technol.* **2004**, *59*, 3–59.

50. Elgersma, A.; Tamminga, S.; Ellen, G. Modifying milk composition through forage. *Anim. Feed Sci. Technol.* **2006**, *131*, 207–225. [CrossRef]

51. Troegeler-Meynadier, A.; Bret-Bennis, L.; Enjalbert, F. Rates and efficiencies of reactions of ruminal biohydrogenation of linoleic acid according to pH and polyunsaturated fatty acids concentrations. *Reprod. Nutr. Dev.* **2006**, *46*, 713–724. [CrossRef] [PubMed]

52. Capuano, E.; Van der Veer, G.; Boerrigter-Eenling, R.; Elgersma, A.; Rademaker, J.; Sterian, A.; Van Ruth, S.M. Verification of fresh grass feeding, pasture grazing and organic farming by cows farm milk fatty acid profile. *Food Chem.* **2014**, *164*, 234–241. [CrossRef] [PubMed]

53. Botana, A.; González, L.; Dagnac, T.; Resch-Zafra, C.; Pereira-Crespo, S.; Veiga, M.; Lorenzana, R. Fatty acids and lipo-soluble antioxidants in milk from dairy farms in the Atlantic area of Spain. *Grassl. Sci. Eur.* **2018**, *23*, 712–714.

54. Butler, G.; Nielsen, J.H.; Slots, T.; Seal, C.; Eyre, M.D.; Sanderson, R.; Leifert, C. Fatty acid and fat soluble antioxidant concentrations in milk from high-and low-input conventional and organic systems: Seasonal variation. *J. Sci. Food Agric.* **2008**, *88*, 1431–1441. [CrossRef]

55. Średnicka-Tober, D.; Barański, M.; Seal, C.J.; Sanderson, R.; Benbrook, C.; Steinshamn, H.; Gromadzka-Ostrowska, J.; Rembiałkowska, E.; Skwarło-Sońta, K.; Eyre, M.; et al. Higher PUFA and n-3 PUFA, conjugated linoleic acid, α-tocopherol and iron, but lower iodine and selenium concentrations in organic milk: A systematic literature review and meta and redundancy analyses. *Br. J. Nutr.* **2016**, *115*, 1043–1060. [CrossRef] [PubMed]

56. Cichosz, G.; Czeczot, H.; Ambroziak, A.; Bielecka, M.M. Natural antioxidants in milk and dairy products. *Int. J. Dairy Technol.* **2017**, *70*, 165–178. [CrossRef]

57. Sunarić, S.; Živković, J.; Pavlović, R.; Kocić, G.; Trutić, N.; Živanović, S. Assessment of α-tocopherol content in cow and goat milk from the Serbian market. *Hemijskaindustrija* **2012**, *66*, 559–566. [CrossRef]

58. Nozière, P.; Graulet, B.; Lucas, A.; Martin, B.; Grolier, P.; Doreau, M. Carotenoids for ruminants: From forages to dairy products. *Anim. Feed Sci. Technol.* **2006**, *131*, 418–450. [CrossRef]

 *foods*

Article

Seasonal Variation of Chemical Composition, Fatty Acid Profile, and Sensory Properties of a Mountain Pecorino Cheese

Francesco Serrapica [1], Felicia Masucci [1,*], Antonio Di Francia [1], Fabio Napolitano [2], Ada Braghieri [2], Giulia Esposito [3,4] and Raffaele Romano [1]

[1] Dipartimento di Agraria, Università di Napoli Federico II, Via Università 100, 80055 Portici, Italy; francesco.serrapica@unina.it (F.S.); antonio.difrancia@unina.it (A.D.F.); raffaele.romano@unina.it (R.R.)

[2] Scuola di Scienze Agrarie, Forestali, Alimentari ed Ambientali, Università degli Studi della Basilicata, Via dell'Ateneo Lucano 10, 85100 Potenza, Italy; fabio.napolitano@unibas.it (F.N.); ada.braghieri@unibas.it (A.B.)

[3] Department of Animal Sciences, Faculty of AgriSciences, Stellenbosch University, Matieland 7602, South Africa; giulia@sun.ac.za

[4] Research and Development RUM & N Sas, via Sant'Ambrogio 4/A, 42123 Reggio Emilia, Italy

* Correspondence: masucci@unina.it; Tel.: +39-081-253-9307

Received: 9 July 2020; Accepted: 7 August 2020; Published: 10 August 2020

Abstract: This study aims to assess the compositional traits and sensory characteristics of a traditional pecorino cheese associated with management and feeding system seasonality. The study was carried out on two mountain dairy farms using an outdoor, pasture-based system from April to October (OutS), and an indoor system (InS) during the rest of the year. Outdoor-produced milk had higher fat content and a tendency for protein and somatic cell count to be higher. The OutS cheeses showed higher dry matter and fat content, higher percentages of unsaturated fatty acids, C18:3, *cis*-9, *trans*-11 conjugated linoleic acid, and *trans*-11 C18:1, and lower percentages of C14:0 and C16:0. These modifications in fatty acid composition determined the reduction of the atherogenic index. The OutS cheeses also displayed higher intensity of almost all sensory attributes, including odor, flavor, taste, and texture descriptors. The outdoor system partly reduced the liking of consumers for pecorino. However, changes in the productive process leading to an increment in the water content and softness of the cheeses (i.e., controlled humidity and temperature during ripening) may increase the overall liking of pasture-based products, thus promoting the consumption of healthier foods.

Keywords: pecorino cheese; pasture; management system; fatty acids profile; sensory properties; consumer liking

---

1. Introduction

Locally produced traditional foods are positively perceived by consumers and, although a role is played by consumer feelings such as nostalgia, ethnocentrism, and need of authenticity, sensory properties remain the main determinant of purchase intent and food liking [1]. The sensory properties of cheese can be affected by several environmental and technological factors [2,3]. In particular, feeding plays a central role in affecting milk characteristics, which in turn shape cheese quality traits [4]. Forages, which represent the main ingredient of ruminants' diets, are able to convey specific nutritional and organoleptic features to milk. Thus, they can significantly contribute to define cheese "terroir" and healthiness [5]. Several studies have focused on the effect of botanical diversity and preservation methods of forage (fresh, hay, or ensiled) on the quality of various cheeses [6–9]. In particular, it has been well established that pasture and fresh herbage can positively change the fatty acid (FA) composition of milk fat in terms of polyunsaturated fatty acids (PUFA). The latter are transferred into the corresponding dairy products affecting cheese flavor and texture [10].

In many non-irrigable inland areas of the European Mediterranean regions, pasture-based sheep farming sustains the socio-economic vitality of local rural communities, also preventing environmental damage and changes in the traditional agricultural landscape [11]. In the regions of southern Italy, the extensive and semi-extensive sheep farms are mainly devoted to produce light lambs (22–24 kg live weight) as well as milk processed in several cheese varieties linked to the local traditions, generically named pecorino. In these areas, the pastures are characterized by a marked seasonal fluctuation of botanical composition and productivity [12]. The vegetative cycle begins in autumn, as the first rains appear, passes through a winter dormancy period imposed by the low temperatures, and then reaches maximum vegetative vigor in the late spring and early summer due to the favorable combinations of temperature and rainfall [12]. Consequently, the farming system seasonally changes from extensive into semi-extensive or indoors and, accordingly, from pasture grazing to diets based on preserved forages. Changes in ewe milk yield and composition associated with seasonal variations of feeding regimes have been addressed by a large body of literature, and the effects on physiochemical characteristics and fatty acid profiles of milk have been reviewed [13,14]. Nevertheless, studies addressing seasonal variation on pecorino cheese characteristics are scanty [15–17], and none of them have examined the impact of feeding regime on cheese sensory and chemical properties.

Thus, this study aimed to assess in Bagnolese pecorino (a cheese recognized as a Traditional Agri-Food Product by the Italian Ministry of Agriculture) the compositional traits and sensory characteristics associated with management and feeding system seasonality. In addition, the consumer liking for cheeses made by different dairies and in different systems was evaluated.

2. Materials and Methods

2.1. Study Site

The study was conducted at two family-owned, small-scale dairy sheep farms (A and B) located in an internal upland area (41°00′ N 15°26′ E and 41°01′ N 15°32′ E; about 900 m above sea level) of Campania, a Region of southern Italy. The farms were selected based on two main criteria: (1) use of pasture as the main feeding source during the grazing season; and (2) Bagnolese pecorino cheese entirely produced from farm milk. Farming was based on a mixed cereals–semi extensive sheep production system. The available dry arable land was used for cropping winter cereals and mixed meadows for hay production. The grazing pastures were private, and the flocks were moved to different grazing areas when forage was considered scarce. Both farms raised the native sheep breed "Bagnolese", averaging three lambings every two years. Lambs were milk-fed until slaughter, at about 30–40 days of age, and ewes were milked thereafter, once a day, at evening. From April to October the flocks were kept on pasture integrated in the case of lactating animals with hay and concentrate mixture. Ewes were reared indoors during the rest of the year, when they were fed hay and cereal grains. Semi-arid grassland providing annual herbaceous plants, shrub pastures and, during drought, summer stubbles were the resources available for grazing. The main farm characteristics are reported in Table 1.

Table 1. Farm characteristics and ingredients and composition of the diets fed in the barn under the two management systems (raw means ± SD).

Item	InS		OutS	
	Farm A	**Farm B**	**Farm A**	**Farm B**
Farm Characteristics				
Usable agricultural area, ha	21	17	21	17
Available grazing areas, ha	-	-	14	8
Lactating ewes, no.	106	78	140	104
Ingredients of the Diets Fed in the Barn				
Concentrate [1], kg of DM/head per day	0.8	0.8	0.25	0.25
Hay [2], kg of DM/head per day	1.8	1.9	0.25	0.40
Composition of the Diets Fed in Barn				
Crude protein, % of DM	14.1 ± 0.7	15.4 ± 0.8	14.9 ± 0.5	15.1 ± 0.7
Ether extract, % of DM	2.5 ± 0.4	1.2 ± 0.6	2.6 ± 0.2	1.1 ± 0.4
NDF, % of DM	46.6 ± 1.4	51.7 ± 1.3	40.7 ± 1.1	48.8 ± 1.8
ADF, % of DM	29.6 ± 1.8	23.6 ± 1.5	23.5 ± 0.9	18.2 ± 0.6
NE_L, MJ/kg of DM	3.6 ± 0.13	3.9 ± 0.17	4.4 ± 0.15	4.5 ± 0.2

[1] Based on barley, fava beans, oat, and wheat meal. [2] Natural meadow and clover hays. InS, indoor system; OutS, outdoor system; DM, dry matter; NDF, neutral detergent fiber; NE_L, net energy of lactation; SD, standard deviation.

2.2. Experimental Design and Sampling Procedure

According to the pasture availability, the management system, and the season, the study was divided into two periods that are designated hereafter as Outdoor (OutS) and Indoor systems (InS). The InS period was from January to February, and the OutS period from April to May. During each period, each farm was visited at 2-week intervals (3 visits/farm/period). In the OutS period, the grass was sampled from movable grazing exclusion cages (1.5 m × 1.5 m) randomly located in the pasture before the grazing season, and in both OutS and InS periods the feeds fed in the barns were collected, and the bulk milk (200 mL) was gathered before cheese making.

The pecorino cheese manufacturing process at the artisanal dairy plants adjoining the farms was followed at each visit. The two dairies used similar cheese-making processes to produce the local semi-cooked cheese "Bagnolese pecorino". Briefly, filtered raw ewe milk was gently heated in a copper vat to 36–38 °C, and kid paste rennet was added. At curd formation, the coagulum was cut by a wooden stick to rice size, cooked at 42 C° for about 5 min, and transferred into perforated plastic molds (about 2 L in volume) and pressed by hand to allow the whey drainage. Molded curd was left to acidify overnight and then dry salted by hand. The pecorino cheeses produced from the sampled milk were marked so as to be identified later. These cheeses were air-ripened in cool farm cellars for 60 d with frequent rotations and cleaning of the surface and then sent to the laboratory for analyses.

2.3. Analysis

2.3.1. Chemical and Instrumental Analysis of Milk and Cheese

The samples of hay ($n = 6$, 2 farms × 3 sampling times) and pasture ($n = 18$, 2 farms × 3 sampling time × 3 cages) were dried in a forced-air oven (at 65 °C until constant weight), ground to pass a 1-mm screen, and then separately analyzed according the Association of Official Analytical Chemists (AOAC) for dry matter (DM), ash, ether extract, and crude protein [18]. The method of Van Soest et al. [19] was used to determine the neutral detergent fiber (NDF) and acid detergent fiber (ADF) contents. Energy content, expressed in MJ of net energy of lactation (NEL), was estimated according to the Sauvant and Nozière equation [20]. The milk fat and protein contents and somatic cell count (SCC) were determined by mid-infrared spectrophotometry (MilkoScan FT 6000 and Fossomatic 90, Foss Electric, Hillerød, Denmark). The chemical and fatty acid (FA) composition of pecorino was determined by sampling (about 300 g) each cheese ($n = 6$, 2 farms × 2 periods × 3 samples) at 1 cm from the rind. Quantification of moisture was performed by oven drying, and the fat and protein contents by the Gerber and Kjeldahl methods, respectively [18]. Fatty acid composition was determined after lipid extraction according the Schmidt–Bondzynski–Ratzlaff method, as reported by Romano et al. [21]. A Supelco 37 Component FAME mix (Supelco, Bellefonte, PA, USA) and a CLA (Conjugated Linoleic Acid) isomer mixture (Nu-Chek Prep. Inc., Elysian, MN, USA) were used as standards. Fatty acid values <0.1 were not quantified and atherogenic index was calculated [22]. Cheese color (CIELAB system) was measured on 3 non overlapping areas of a slice of cheese (Minolta colorimeter CR-300, Minolta Camera Co. Ltd., Osaka, Japan), as described elsewhere [23].

2.3.2. Sensory Analysis and Consumer Liking of Pecorino Cheese

A quantitative–descriptive analysis (QDA) method [24] was used to assess pecorino cheese sensory properties. Sixteen subjects chosen among regular eaters of pecorino cheese (they had to consume pecorino cheese at least twice a month) were recruited by telephone. Among them, twelve panelists (average age 25 years) were selected following the recommendations issued by International Organization for Standardization (ISO) [25]. In particular, the sensitivity of the subjects to the four basic tastes was used [26]. During four preliminary sessions, the assessors developed a consensus list of attributes based on the available literature [27,28], resulting in a single score card with 2 odor, 2 flavor, 5 taste, and 5 texture descriptors (Table 2).

Subsequently, panelists were trained in the assessment of the intensity of sensory stimuli as suggested by Braghieri et al. [29]. Namely, the panel leader proposed a number of different products to be used as standard references for each attribute, and panelists identified which of them were the most appropriate. The completion of the entire reference frame, as reported in Table 2, required two 2 h sessions. In a further 2 h session, assessors were instructed to use the rating scale ranging from 0 (lack of sensation) to 100 (high intensity). For training purposes, at least two points of the scale were anchored using the previously identified references. In particular, panelists were informed about the intensity level of the samples to which they were exposed for three times. Then, they had to identify the intensity level of each attribute without any information. For this training eight 2 h sessions were required.

As to the actual QDA, in each session test, carried out at about 10.00 h, three rind-free cheese cubes (1 cm^3) were served in random order and panelists gave a score between 0 and 100 for the intensity of each attribute. The panelists assessed five replications of each sample in sensory booths. In all, 5 sessions were conducted, and in each session two cheese cubes were offered in monadic randomized order and evaluated by each panelist when the temperature of each sample was 15 °C.

Pecorino cheeses were also evaluated for consumer liking [30] by 100 consumers balanced for sex with an average age of 40 years. Each participant evaluated four rind-free 1 cm^3 cheese cubes (2 seasons by 2 replications). For each sample, consumers expressed an overall liking and a liking according to the following sensory inputs: appearance, texture, and taste/flavor. Consumers rated their liking on a hedonic scale ranging from "extremely unpleasant" (1) to "extremely pleasant" (9) with a central point (5) matching "neither unpleasant nor pleasant" [30].

Table 2. List of attributes, definitions, and reference frame used by a 12-member panel for Bagnolese pecorino cheese sensory profiling.

Descriptor	Definition	Reference Samples	
		Lower Anchor	Upper Anchor
Odor			
Barn	Odor arising from a sheep barn	60 g ricotta cheese	60 g ricotta cheese + 60 g grated pecorino cheese
Hay	Odor arising from hay	200 mL water	5 g hay in 200 mL water
Flavor			
Pecorino	Typical flavor of pecorino cheese	cacioricotta cheese	Bagnolese pecorino cheese
Barn	Flavor arising from a sheep barn	60 g ricotta cheese	60 g ricotta cheese + 60 g grated pecorino cheese
Taste			
Sweet	Taste elicited by sucrose	8 mL stock solution 100 mL^{-1}	20 mL stock solution 100 mL^{-1}
Salty	Taste elicited by sodium chloride	1.5 mL stock solution 100 mL^{-1}	3 mL stock solution 100 mL^{-1}
Acid	Taste elicited by citric acid	8 mL stock solution 10^1	16 mL stock solution 100 mL^{-1}
Bitter	Taste elicited by quinine	4 mL stock solution 100 mL^{-1}	8 mL stock solution 100 mL^{-1}
Spicy	Taste associated with an irritating or aggressive sensation perceived in the mouth or in the throat	10 g ricotta cheese	10 g ricotta cheese + 0.2 g hot pepper in powder
Texture			
Hardness	Highest force required to chew cheese samples	20 g mozzarella cheese	20 g Pecorino cheese ripened 12 months
Friability	Increasing perception of cheese fragments during mastication	20 g Emmental cheese	20 g parmesan cheese ripened 36 months
Grainess	Perception of course particles in the mouth	20 g Fontina cheese	20 g parmesan cheese ripened 36 months
Solubility	Perception of cheese melting in the mouth	20 g Fontina cheese	20 g mini cheese spread
Adhesivity	Effort needed to remove the layer of cheese coating the mouth	20 g mozzarella cheese	20 g Taleggio cheese

2.4. Statistical Analysis

The SAS statistical software (SAS Institute Inc., Cary, NC, USA) was used to perform data analysis. Data on milk and cheese composition were analyzed by two-way analysis of variance (GLM procedure) to determine the fixed effect of management system (OutS and InS), farm (A and B), and their interaction (management system by farm). The sensory attributes of pecorino cheese were analyzed by ANOVA with assessor (12), replication (5), product (4 = 2 management system × 2 farms), and their interactions as factors. In addition, an analysis of variance with management system (2), farm (2), and their interaction as factors was conducted. Similarly, consumer liking was subjected to analysis of variance using management system (2), farm (2), and their interaction as factors.

3. Results and Discussion

3.1. Pasture Composition

The grazing pastures of both farms were mostly composed of Poaceae and Fabaceae with reduced incidence of other grazing species (Table 3), since they were sown at 3–4 year intervals with the grass–legume mixtures as usually managed in the investigated area [31,32].

Table 3. Botanical and chemical compositions of pasture samples (raw means ± SD).

Item	Farm	
	A	**B**
Botanical Composition		
Nonedible biomass,[1] % DM	12.8 ± 1.2	11.5 ± 1.3
Poaceae, % DM of edible biomass	64.5 ± 3.9	63.8 ± 1.3
Lolium spp., % DM of Poaceae	70.8 ± 2.6	65.1 ± 0.6
Other Poaceae,[2] % DM of Poaceae	29.2 ± 2.6	34.9 ± 0.6
Fabaceae, % DM of edible biomass	22.8 ± 4.2	19.2 ± 4.1
Trifolium spp., % DM of Fabaceae	71.1 ± 4.3	61.3 ± 3.8
Other *Fabaceae*,[3] % DM of Fabaceae	28.9 ± 4.3	38.7 ± 3.8
Other species,[4] % DM of edible biomass	12.7 ± 1.1	16.9 ± 3.0
Chemical Composition		
DM, %	23.8 ± 1.3	22.6 ± 0.6
Crude Protein, % of DM	17.5 ± 0.5	17.2 ± 0.5
Ether extract, % of DM	2.3 ± 0.1	2.5 ± 0.1
NDF, % of DM	47.8 ± 2.0	47.2 ± 1.1
ADF, % of DM	22.3 ± 1.3	25.7 ± 3.0
NE_L, MJ/kg of DM	5.4 ± 0.5	5.4 ± 0.2

[1] Nonedible species, mature plants, and dead materials. [2] *Bromus* spp., *Arrhenatherum elatius*, *Avena* spp., *Dactylis glomerata*, *Hordeum* spp., *Festuca* spp., *Agrostis* spp., *Phleum* spp., *Poa* spp., *Agropyron* spp., and *Trisetum* spp. [3] *Medicago* spp., *Lotus corniculatus*, *Vicia* spp., *Hedysarum coronarium*, *Onobrychis viciifolia*, and *Anthyllis vulneraria*. [4] *Carduus* spp., *Cichorium* spp., *Calendula arvensis*, *Chrysanthemum coronarium*, *Mentha* spp., *Achillea* spp., *Borago officinalis*, *Taraxacum* spp., *Foeniculum vulgare*, *Cirsium* spp., and *Plantago* spp. DM, dry matter; NDF, neutral detergent fiber; ADF, acid detergent fiber; NE_L, net energy of lactation; SD, standard deviation.

3.2. Milk Traits and Cheese Chemical Composition and Color

The chemical composition of milk and cheese produced in the two farms as influenced by management system are given in Table 4. In agreement with previous studies [16,33], the OutS milk showed higher ($p < 0.05$) fat content, and a tendency ($p < 0.1$) for protein and SCC to be higher. Due to the market demand for young lambs at Christmas [34,35], in both flocks most of the ewes lambed in autumn, and only primiparous and a small percentage of pluriparous animals (about 20%) lambed at the end of winter. Thus, the observed increments of milk components may be due to the concentration effect related to the lower yield at the end of lactation [16,36]. Moreover, also the progressive deterioration of the ewe udder health in late lactation can increase SCC in ewe milk [37,38].

The proximate chemical composition of pecorino cheese was within the standards required for similar ewe cheeses endowed by protected denomination origin, such as Canestrato Pugliese cheese and the related production rules (e.g., fat content not less than 38% in DM). The cheese OutS showed a higher DM content ($p < 0.01$), indicating a faster loss of moisture during the aging period in summer, and a higher fat percentage ($p < 0.001$), likely related to the tendency observed in OutS milk fat to be higher. Except for the ash content, significant effects of the farm were observed for all the chemical characteristics of pecorino cheese. These differences are expected for traditional cheese produced at small scale dairies [16]. Unsurprisingly, neither farm nor management system affected cheese color. Previous studies reported changes in cheese instrumental color as a consequence of using fresh forage and the closely related β-carotene content [31,39–41]. However, β-carotene is not detectable in ewes' milk due to its enzyme-catalyzed cleavage to retinal in the sheep liver [42]. As a consequence, pecorino cheese color may be unaffected by changes in feeding and β-carotene intake.

Table 4. Milk and cheese chemical composition and color characteristics of cheese (LSM ± SEM) as affected by the management system, farms, and their interaction.

Item	OutS		InS		SEM	Significance [1]		
	Farm A	Farm B	Farm A	Farm B		Management	Farm	M × F
Milk								
Fat, %	6.87	6.92	6.62	7.03	0.10	*	NS	NS
Protein, %	4.80	4.99	4.87	5.21	0.13	+	NS	NS
Lactose, mg/dL	4.80	5.03	4.70	4.73	0.08	NS	*	NS
SCC, $\log_{10}$ *n*. cells/mL	5.65	6.29	5.84	6.02	0.19	+	NS	NS
Cheese								
DM, %	69.73	72.38	67.43	68.6	0.82	**	*	NS
Ash, % of DM	4.61	4.51	4.31	4.51	0.20	NS	NS	NS
Fat, % of DM	49.55	46.58	46.58	44.35	0.56	***	***	NS
Protein, % of DM	32.18	36.08	33.45	37.30	1.17	NS	**	NS
Cheese Color								
L *	72.33	71.88	71.75	70.95	1.26	NS	NS	NS
A *	4.28	5.00	4.60	4.00	0.37	NS	NS	NS
B *	11.10	10.50	11.63	11.08	0.45	NS	NS	NS

[1] * $p \leq 0.05$; ** $p \leq 0.01$ ***; $p \leq 0.001$; + indicates $p \leq 0.10$; NS (not significant) indicates $p > 0.10$. OutS, outdoor system; InS, indoor system; M, management; F, farm; DM, dry matter; SCC, somatic cell count; L *, lightness; A *, red-green color; B *, yellow–blue color; LSM, least square mean; SEM, standard error of mean.

Fatty acid composition of pecorino cheese is given in Table 5. As to the neo-formed (C ≤ 15) and mixed (C16:0 and C16:1) FA [43], the OutS cheeses showed lower values of C12:0 ($p < 0.05$) and C16:0, C14:0 ($p < 0.001$), whereas the C4:0 content was tendentially higher ($p < 0.1$). A reduced synthesis of short and medium-chain FA in the mammary gland may be observed in lactating ruminants on pasture-based diets, due to either high levels of dietary PUFA from fresh forage that compete with de novo FA for the esterification in the mammary gland, or the negative energy balance that may often occur when milk yield is high [10]. Nevertheless, a lack of a marked reduction of de novo FA in animals fed fresh forage has been also reported [31]. In the current work, the low nutrient requirements of the ewes in late stage of lactation might explain the lack of the effect of management system on a number of de novo FA (i.e., C6:0, C8:0, C10:0; C14:1, C15:0). In agreement with numerous authors [43–45], OutS cheeses also showed higher levels of the dietary origin of FAs C18:1 n9*cis*, C18:2 n6 *cis*, C18:1 *trans* 11, ($p < 0.001$), C18:3 n3 ($p < 0.01$), *cis*-9,*trans* 11 conjugated linoleic acid (CLA), and C18:1 *trans*-9 ($p < 0.05$). As an overall result, it was possible to recognize the effect of grazing on cheese FA composition in terms of increased content of PUFA ($p < 0.001$), *cis*-9,*trans*-11 CLA, and *trans*-11 C18:1 along with a reduction of SFA and atherogenic index ($p < 0.001$) under real flock management conditions.

Table 5. Pecorino cheese fatty acids composition as affected by the management system, the farms, and their interaction (LSM ± SEM).

FA, g/100 g of FA	OutS		InS		SEM	Significance [1]		
	Farm A	Farm B	Farm A	Farm B		Management	Farm	M × F
C4:0	5.78	4.04	3.97	3.20	0.61	+	NS	NS
C6:0	3.18	2.79	3.25	3.20	0.32	NS	NS	NS
C8:0	2.66	2.52	2.59	2.27	0.10	NS	*	NS
C10:0	6.58	7.41	7.41	7.25	0.16	NS	NS	*
C12:0	3.57	4.22	4.72	4.82	0.19	*	NS	NS
C14:0	11.39	12.09	13.22	13.79	0.12	***	***	NS
C14:1	1.14	1.20	1.19	1.08	0.07	NS	NS	NS
C15:0	1.43	1.43	1.47	1.42	0.06	NS	NS	NS
C16:0	23.58	23.92	27.37	27.60	0.74	***	***	NS
C16:1	1.08	1.29	1.08	1.25	0.05	NS	**	NS
C17:0	0.67	0.78	0.80	0.76	0.06	NS	NS	NS
C17:1	0.42	0.43	0.42	0.39	0.03	NS	NS	NS
C18:0	10.01	10.59	10.66	10.69	0.64	NS	NS	NS
C18:1 n9 *trans*	4.03	3.50	2.96	2.90	0.31	*	NS	NS
C18:1 *trans*-11	0.38	0.35	0.20	0.15	0.02	***	NS	NS
C18: 1 n9 *cis*	17.57	17.36	14.12	15.02	0.47	***	NS	NS
C18:2 n6	1.83	1.72	1.14	0.99	0.08	***	NS	NS
C18:3 n3	1.57	1.31	0.76	0.77	0.17	**	NS	NS
C20:1	0.25	0.23	0.23	0.20	0.02	NS	NS	NS
Cis-9 *trans*-11 CLA	2.10	1.93	1.60	1.43	0.19	*	NS	NS
C20:4	0.13	0.12	0.15	0.11	0.01	NS	**	NS
Others	0.66	0.76	0.70	0.71	0.03	NS	*	NS
MUFA	24.87	24.40	20.17	21.00	0.43	***	NS	NS
PUFA	5.64	5.08	3.66	3.31	0.23	*	NS	NS
SFA	68.84	69.79	75.45	74.99	0.42	***	NS	NS
Atherogenic index [2]	2.38	2.60	3.57	3.60	0.08	***	NS	NS

[1] * $p \leq 0.05$; ** $p \leq 0.01$ ***; $p \leq 0.001$; + indicates $p \leq 0.10$; NS (not significant) indicates $p > 0.10$. [2] (C12:0 + (4 × C14:0) + C16:0)/unsaturated FA. OutS, outdoor system; InS, indoor system; M, management; F, farm; FA, fatty acids; MUFA, monounsaturated fatty acids; PUFA, polyunsaturated fatty acids; SFA, saturated fatty acids; LSM, least square mean; SEM, standard error of mean.

A large body of literature (Mele [46], Nudda et al. [13], and Cabiddu et al. [14]) reports a "seasonal" effect on the FA composition of the milk fat, with an increase of the concentration of polyunsaturated PUFA and CLA in spring, due to the corresponding increment in the content in polyunsaturated FA in fresh forage, including C18:3. In addition, the highly degradable nitrogen and fiber of fresh forage along with a higher rumen pH related to grazing activity may promote the rumen bacteria producing CLA *cis*-9,*trans*-11 while inhibiting the biohydrogenation of C18:1 *trans*-11, which is converted to CLA *cis*-9,*trans*-11 via Δ9-desaturase in the mammary gland [47]. Although several studies investigated FA composition of ewe milk fat as influenced by the season of production, the literature on pecorino cheese is limited [15–17,48–51]. The use of vegetable seeds and oils [52–57] led to changes in fatty acid composition of ewe milk and cheese higher than those we observed. According to Dewhurst et al. [47] pasture is less efficient than oil or concentrates for modifying milk FA, but, on the other hand, it does not impact on feeding costs, and contributes to the basis for the "terroir" notion [6]. In addition, consumers are interested in traditional cheese with healthier FA profile, but without a substantial increase of the purchase price [58]. The farm of production slightly influenced the acidic composition of the cheese. This result can be explained based on the similar winter-feeding regimen (i.e., hays and grain meals) and similar pasture composition (i.e., semi-natural sown pasture) of the two farms, according with the results of other on-farm studies reporting that differences in milk FA composition are mainly related to the diet and pasture botanical composition [31,59–61].

3.3. Cheese Sensory Properties and Consumer Liking

An important outcome to emerge from this study is represented by the lack of any significant interactions of product with both replication and assessor. This finding suggests that the panel was subjected to an effective training as the products were consistently evaluated across replications and assessors. Based on these preliminary results, we conducted a second analysis of variance with farm, management system, and their interaction as factors. The second analysis showed that several sensory properties of pecorino cheese were affected by the management system (Table 6) as only the taste attributes "acid" and "bitter" and the texture attribute "solubility" were unaffected. Overall, the OutS cheeses showed a higher intensity of almost all sensory attributes. In particular, the trained panel perceived a higher intensity of the odor attributes "barn" ($p < 0.001$) and "hay" ($p < 0.05$), as well as a higher intensity of the flavor attributes "barn" ($p < 0.01$) and "pecorino" ($p < 0.001$) in OutS cheeses than InS. OutS cheeses were also perceived as more "salty" ($p < 0.001$) and "spicy" ($p < 0.05$) and less "sweet" ($p < 0.001$) than InS in terms of taste attributes. In addition, OutS cheeses showed higher intensities of the texture attributes "hardness", "friability", and "graininess" ($p < 0.001$), whereas they had a lower intensity of adhesivity ($p < 0.05$). The literature regarding cheese produced from animals on pasture does not provide unequivocal indications as far as sensory properties are concerned, since the cheese making procedure, including the use of starter cultures and the ripening conditions, can markedly influence the sensory properties of cheeses [62,63], thus possibly confounding the effects of diet on milk components. However, when the cheese-making process is conducted under controlled conditions, some cheese sensory properties may be traced back to the diet [6]. Pasture is rich in odor-active compounds that can be transferred to cheese [39]. In addition, the higher amount of MUFA and PUFA produced in pasture-based systems can influence the development of compounds which are active in terms of flavor and odor, and that is particularly so in aged cheese [64]. Nevertheless, conflicting results are reported on the effect of grazing on odor/flavor attributes. In agreement with our results, Cantal cheese produced from grazing pasture showed higher intensities of overall odor and aroma compared with the indoor system [65]. Conversely, a lower intensity of odor as a consequence of the ingestion of fresh forage was observed in ripened and fresh pasta filata cheeses [7,31]. These contrasting results may likely to be due to differences in the cheese making process. Both Bagnolese pecorino and Cantal are semi-cooked cheeses produced from raw milk so that neither milk nor curd are subjected to temperatures higher than 50 °C. By contrast, during pasta filata cheese making, the curd is stretched in water or whey heated at 80–90 °C. This thermal treatment may at least partly reduce the activity of the odor compounds in the milk obtained from grazing animals thus flattening the odor profile of the corresponding cheese. Accordingly, it has been recently observed that the amount of volatile organic compounds of milk may be markedly reduced in mozzarella, due to the high temperatures of curd during stretching [66].

In agreement with McSweeney [67], in summer a higher ripening temperature may have induced higher proteolysis levels with the production of amino acids and short peptides responsible for a higher intensity of the attribute "spicy" in OutS cheeses. Similarly, in the same products (i.e., OutS cheeses) a higher water loss related to a higher ripening temperature may have also caused an increase in the intensity of the taste attribute "salty" and a corresponding reduction of the intensity of the attribute "sweetness". As to texture attributes, higher intensity values of "hardness" were unexpectedly perceived in OutS cheese as compared with InS cheese. As also reported in previous studies, a profile richer in unsaturated FAs, which are characterized by a lower melting point, may lead to softer and spreadable cheeses [6,7]. However, our results are not necessarily in contrast with those reported in previous studies. They may be rather attributed to the lower moisture content observed in OutS cheeses, which in turn can be due to the higher ripening temperature occurring in summer. The higher values of friability and graininess observed in OutS cheeses may be also attributed to the effect of the ripening temperature, as a lower water content may have caused a more intense perception of course particles in the mouth (i.e., "grainess"), including the formation of new particles (i.e., "friability"), and a lower perception of cheese adhering to the mouth during mastication (i.e., "adhesivity").

Table 6. Sensory profile, as assessed by a 12-member trained panel, and hedonic scores, as scored by a 100-member untrained consumer panel, of Bagnolese pecorino cheese affected by the management system, the farms, and their interaction (LSM ± SEM).

Descriptor	OutS		InS		SEM	Significance [1]		
	Farm A	Farm B	Farm A	Farm B		Management	Farm	M × F
Odour								
Barn	36.38	72.91	33.36	50.11	3.9	**	***	*
Hay	36.20	49.44	34.84	35.09	4.0	*	*	*
Flavor								
Pecorino	43.04	65.07	35.66	51.15	3.7	**	***	NS
Barn	23.18	58.88	24.16	38.89	3.4	**	***	**
Taste								
Sweet	7.73	3.27	30.86	14.67	2.5	***	***	*
Salty	35.51	58.36	18.39	39.26	3.0	***	***	NS
Acid	11.96	26.49	16.93	22.41	3.0	NS	**	NS
Bitter	14.29	24.53	13.12	23.22	3.1	NS	**	NS
Spicy	14.38	31.53	10.18	19.65	3.1	*	***	NS
Texture								
Hardness	55.33	71.82	21.65	43.85	2.7	***	***	NS
Friability	44.69	55.04	25.02	44.82	3.5	***	***	NS
Grainess	51.18	68.20	23.39	46.33	3.5	***	***	NS
Solubility	23.11	22.04	30.27	25.61	3.5	NS	NS	NS
Adhesivity	26.20	25.67	32.52	33.82	3.4	*	NS	NS
Hedonic scores								
Overall liking	6.44	6.13	7.23	7.06	0.21	***	NS	NS
Appearance	6.63	6.18	7.35	7.07	0.19	***	*	NS
Taste/flavor	6.33	6.16	7.03	7.00	0.21	***	NS	NS
Texture	6.53	6.04	7.34	6.89	0.20	***	*	NS

[1] * $p \leq 0.05$; ** $p \leq 0.01$ ***; $p \leq 0.001$; NS (not significant) indicates $p > 0.10$. OutS, outdoor system; InS, indoor system; M, management; F, farm; LSM, least square mean; SEM, standard error of mean.

Both products were well received by consumers as they received scores higher than the central point for overall liking as well as for the liking of appearance, odor/flavor, and texture (Table 6). Consumers were able to distinguish OutS cheeses (i.e., made using milk of grazing animals) from InS cheeses (i.e., made using milk of animals kept indoors). In particular, they expressed higher levels of liking (i.e., overall, appearance, texture, and odor/flavor) for InS cheeses as compared with OutS ($p < 0.001$). In addition, the farm affected the liking of appearance and texture ($p < 0.05$; data not shown). Texture and flavor attributes represent important sets of sensory properties markedly affecting cheese overall quality and consumer liking. Previous studies showed that consumers may express higher levels of liking for products showing characteristics related to freshness [68] with a preference for cheeses with higher perceived moisture and tenderness intensities [62]. Therefore, the higher ripening temperature in summer may have played a negative role in affecting consumer liking for cheeses obtained from grazing animals, and the use of aging facilities with controlled humidity and temperature may be suggested in order to increase the acceptability of pasture-based products.

4. Conclusions

The main result of the present study is that the pasture-based, outdoor management system has improved health characteristics of cheese in terms of increased content of PUFA, *cis*-9, *trans*-11 CLA, and *trans*-11 C18:1, along with a reduction of SFA and atherogenic index. Overall, the cheeses obtained from the OutS system showed a higher intensity of almost all sensory attributes, including odor, flavor, taste, and texture descriptors. Concomitantly although all cheeses were scored well above the central point, pasture grazing partly reduced the liking of consumers for the pecorino.

However, changes in the productive process leading to an increment in the water content and softness of the cheeses (i.e., controlled humidity and temperature conditions during ripening) may increase the overall liking of pasture-based products in order to meet the sensory preferences of a higher number of consumers and promote the consumption of healthier foods. The farm of production marginally influenced some of the chemical and sensory characteristics of the cheeses. Overall, the healthier nutritional characteristics, if paired with appropriate sensory characteristics, may strengthen the identity of the mountain cheese while sustaining the income of local people based on upland farming.

Author Contributions: Conceptualization, F.M., A.D.F. and F.S.; methodology, F.M., F.N., A.B. and R.R.; formal analysis, G.E. and A.B.; resources, G.E.; data curation, F.M., F.S. and F.N.; writing—original draft, F.S., F.M. and F.N.; writing—review and editing, F.S., F.M. and F.N.; supervision, R.R.; funding acquisition, A.D.F. All authors have read and agreed to the published version of the manuscript.

Funding: This research was funded by REGIONE CAMPANIA, D.R.D no. 443 06/12/2007.

Acknowledgments: The authors would like to thank Amelia Riviezzi (Scuola di Scienze Agrarie, Forestali, Alimentari ed Ambientali, Università degli Studi della Basilicata, Potenza, Italy) for technical assistance in sensory analysis, and Roberto Di Matteo (Dipartimento di Agraria, Università di Napoli "Federico II", Portici, Italy) for cheese color analysis.

Conflicts of Interest: The authors declare no conflict of interest.

References

1. Braghieri, A.; Girolami, A.; Riviezzi, A.M.; Piazzolla, N.; Napolitano, F. Liking of Traditional Cheese and Consumer Willingness to Pay. *Ital. J. Anim. Sci.* **2014**, *13*, 3029. [CrossRef]

2. Uzun, P.; Serrapica, F.; Masucci, F.; Barone, C.M.A.; Yildiz, H.; Grasso, F.; Di Francia, A. Diversity of traditional Caciocavallo cheeses produced in Italy. *Int. J. Dairy Technol.* **2020**, *73*, 234–243. [CrossRef]

3. Fox, P.F.; Cogan, T.M.; Guinee, T.P. Chapter 25—Factors That Affect the Quality of Cheese. In *Cheese*, 4th ed.; McSweeney, P.L.H., Fox, P.F., Cotter, P.D., Everett, D.W., Eds.; Academic Press: San Diego, CA, USA, 2017; pp. 617–641. ISBN 978-0-12-417012-4.

4. Sabia, E.; Gauly, M.; Napolitano, F.; Cifuni, G.F.; Claps, S. The effect of different dietary treatments on volatile organic compounds and aromatic characteristics of buffalo Mozzarella cheese. *Int. J. Dairy Technol.* **2020**, *73*, 594–603. [CrossRef]

5. Giaccone, D.; Revello-Chion, A.; Galassi, L.; Bianchi, P.; Battelli, G.; Coppa, M.; Tabacco, E.; Borreani, G. Effect of milk thermisation and farming system on cheese sensory profile and fatty acid composition. *Int. Dairy J.* **2016**, *59*, 10–19. [CrossRef]

6. Martin, B.; Verdier-Metz, I.; Buchin, S.; Hurtaud, C.; Coulon, J.-B. How do the nature of forages and pasture diversity influence the sensory quality of dairy livestock products? *Anim. Sci.* **2005**, *81*, 205–212. [CrossRef]

7. Uzun, P.; Masucci, F.; Serrapica, F.; Napolitano, F.; Braghieri, A.; Romano, R.; Manzo, N.; Esposito, G.; Di Francia, A. The inclusion of fresh forage in the lactating buffalo diet affects fatty acid and sensory profile of mozzarella cheese. *J. Dairy Sci.* **2018**, *101*, 6752–6761. [CrossRef]

8. Giorgio, D.; Di Trana, A.; Di Napoli, M.; Sepe, L.; Cecchini, S.; Rossi, R.; Claps, S. Comparison of cheeses from goats fed 7 forages based on a new health index. *J. Dairy Sci.* **2019**, *102*, 6790–6801. [CrossRef] [PubMed]

9. Serrapica, F.; Uzun, P.; Masucci, F.; Napolitano, F.; Braghieri, A.; Genovese, A.; Sacchi, R.; Romano, R.; Barone, C.M.A.; Di Francia, A. Hay or silage? How the forage preservation method changes the volatile compounds and sensory properties of Caciocavallo cheese. *J. Dairy Sci.* **2020**, *103*, 1391–1403. [CrossRef]

10. Alothman, M.; Hogan, S.A.; Hennessy, D.; Dillon, P.; Kilcawley, K.N.; O'Donovan, M.; Tobin, J.T.; Fenelon, M.; O'Callaghan, T.F. The "Grass-Fed" Milk Story: Understanding the Impact of Pasture Feeding on the Composition and Quality of Bovine Milk. *Foods* **2019**, *8*, 350. [CrossRef]

11. Sabia, E.; Gauly, M.; Napolitano, F.; Serrapica, F.; Cifuni, G.F.; Claps, S. Dairy sheep carbon footprint and ReCiPe end-point study. *Small Rumin. Res.* **2020**, *185*, 106085. [CrossRef]

12. Scocco, P.; Piermarteri, K.; Malfatti, A.; Tardella, F.M.; Catorci, A. Increase of drought stress negatively affects the sustainability of extensive sheep farming in sub-Mediterranean climate. *J. Arid. Environ.* **2016**, *128*, 50–58. [CrossRef]

13. Nudda, A.; Battacone, G.; Neto, O.B.; Cannas, A.; Francesconi, A.H.D.; Atzori, A.; Pulina, G. Feeding strategies to design the fatty acid profile of sheep milk and cheese. *Rev. Bras. de Zootec.* **2014**, *43*, 445–456. [CrossRef]

14. Cabiddu, A.; Delgadillo-Puga, C.; DeCandia, M.; Molle, G.; Delgadillo-Puga, C. Extensive Ruminant Production Systems and Milk Quality with Emphasis on Unsaturated Fatty Acids, Volatile Compounds, Antioxidant Protection Degree and Phenol Content. *Animals* **2019**, *9*, 771. [CrossRef] [PubMed]

15. Nudda, A.; McGuire, M.; Battacone, G.; Pulina, G. Seasonal Variation in Conjugated Linoleic Acid and Vaccenic Acid in Milk Fat of Sheep and its Transfer to Cheese and Ricotta. *J. Dairy Sci.* **2005**, *88*, 1311–1319. [CrossRef]

16. Todaro, M.; Bonanno, A.; Scatassa, M.L. The quality of Valle del Belice sheep's milk and cheese produced in the hot summer season in Sicily. *Dairy Sci. Technol.* **2013**, *94*, 225–239. [CrossRef]

17. Addis, M.; Fiori, M.; Riu, G.; Pes, M.; Salvatore, E.; Pirisi, A. Physico-chemical characteristics and acidic profile of PDO Pecorino Romano cheese: Seasonal variation. *Small Rumin. Res.* **2015**, *126*, 73–79. [CrossRef]

18. Association of Official Analytical Chemists (AOAC). *Official Methods of Analysis*, 17th ed.; Horwitz, W., Ed.; AOAC International: Gaithersburg, MD, USA, 2002.

19. Van Soest, P.; Robertson, J.; Lewis, B. Methods for Dietary Fiber, Neutral Detergent Fiber, and Nonstarch Polysaccharides in Relation to Animal Nutrition. *J. Dairy Sci.* **1991**, *74*, 3583–3597. [CrossRef]

20. Sauvant, D.; Nozière, P. La quantification des principaux phénomènes digestifs chez les ruminants: Les relations utilisées pour rénover les systèmes d'unités d'alimentation énergétique et protéique. *INRA Prod. Anim.* **2013**, *26*, 327–346. [CrossRef]

21. Romano, R.; Masucci, F.; Giordano, A.; Spagna Musso, S.; Naviglio, D.; Santini, A. Effect of tomato by-products in the diet of Comisana sheep on composition and conjugated linoleic acid content of milk fat. *Int. Dairy J.* **2010**, *20*, 858–862. [CrossRef]

22. Ulbricht, T.; Southgate, D. Coronary heart disease: Seven dietary factors. *Lancet* **1991**, *338*, 985–992. [CrossRef]

23. Masucci, F.; De Rosa, G.; Barone, C.M.A.; Napolitano, F.; Grasso, F.; Uzun, P.; Di Francia, A. Effect of group size and maize silage dietary levels on behaviour, health, carcass and meat quality of Mediterranean buffaloes. *Animals* **2015**, *10*, 531–538. [CrossRef] [PubMed]

24. Murray, J.; Delahunty, C.; Baxter, I. Descriptive sensory analysis: Past, present and future. *Food Res. Int.* **2001**, *34*, 461–471. [CrossRef]

25. ISO. *ISO 8586:2012 Sensory Analysis—General Guidelines for the Selection, Training and Monitoring of Selected Assessors and Expert Sensory Assessor*; International Organization for Standardization: Geneva, Switzerland, 2012; Available online: https://www.iso.org/obp/ui/#iso:std:iso:8586:ed-1:v2:en (accessed on 7 July 2020).

26. Albenzio, M.; Santillo, A.; Caroprese, M.; Braghieri, A.; Sevi, A.; Napolitano, F. Composition and sensory profiling of probiotic Scamorza ewe milk cheese. *J. Dairy Sci.* **2013**, *96*, 2792–2800. [CrossRef] [PubMed]

27. Di Cagno, R.; Banks, J.; Sheehan, L.; Fox, P.F.; Brechany, E.; Corsetti, A.; Gobbetti, M. Comparison of the microbiological, compositional, biochemical, volatile profile and sensory characteristics of three Italian PDO ewes' milk cheeses. *Int. Dairy J.* **2003**, *13*, 961–972. [CrossRef]

28. Torracca, B.; Pedonese, F.; López, M.B.; Turchi, B.; Fratini, F.; Nuvoloni, R. Effect of milk pasteurisation and of ripening in a cave on biogenic amine content and sensory properties of a pecorino cheese. *Int. Dairy J.* **2016**, *61*, 189–195. [CrossRef]

29. Braghieri, A.; Zotta, T.; Morone, G.; Piazzolla, N.; Majlesi, M.; Napolitano, F. Starter cultures and preservation liquids modulate consumer liking and shelf life of mozzarella cheese. *Int. Dairy J.* **2018**, *85*, 254–262. [CrossRef]

30. Markey, O.; Souroullas, K.; Fagan, C.C.; Kliem, K.E.; Vasilopoulou, D.; Jackson, K.G.; Humphries, D.J.; Grandison, A.S.; Givens, D.I.; Lovegrove, J.A.; et al. Consumer acceptance of dairy products with a saturated fatty acid–reduced, monounsaturated fatty acid–enriched content. *J. Dairy Sci.* **2017**, *100*, 7953–7966. [CrossRef]

31. Esposito, G.; Masucci, F.; Napolitano, F.; Braghieri, A.; Romano, R.; Manzo, N.; Di Francia, A. Fatty acid and sensory profiles of Caciocavallo cheese as affected by management system. *J. Dairy Sci.* **2014**, *97*, 1918–1928. [CrossRef]

32. Uzun, P.; Masucci, F.; Serrapica, F.; Varricchio, M.L.; Pacelli, C.; Claps, S.; Di Francia, A. Use of mycorrhizal inoculum under low fertilizer application: Effects on forage yield, milk production, and energetic and economic efficiency. *J. Agric. Sci.* **2018**, *156*, 127–135. [CrossRef]

33. Sevi, A.; Albenzio, M.; Marino, R.; Santillo, A.; Muscio, A. Effects of lambing season and stage of lactation on ewe milk quality. *Small Rumin. Res.* **2004**, *51*, 251–259. [CrossRef]

34. Sitzia, M.; Bonanno, A.; Todaro, M.; Cannas, A.; Atzori, A.S.; Francesconi, A.H.D.; Trabalza-Marinucci, M. Feeding and management techniques to favour summer sheep milk and cheese production in the Mediterranean environment. *Small Rumin. Res.* **2015**, *126*, 43–58. [CrossRef]

35. Todaro, M.; Dattena, M.; Acciaioli, A.; Bonanno, A.; Bruni, G.; Caroprese, M.; Mele, M.; Sevi, A.C.; Trabalza-Marinucci, M. Aseasonal sheep and goat milk production in the Mediterranean area: Physiological and technical insights. *Small Rumin. Res.* **2015**, *126*, 59–66. [CrossRef]

36. Bianchi, L.; Casoli, C.; Pauselli, M.; Budelli, E.; Caroli, A.; Bolla, A.; Duranti, E. Effect of somatic cell count and lactation stage on sheep milk quality. *Ital. J. Anim. Sci.* **2004**, *3*, 147–156. [CrossRef]

37. Fthenakis, G. Prevalence and aetiology of subclinical mastitis in ewes of Southern Greece. *Small Rumin. Res.* **1994**, *13*, 293–300. [CrossRef]

38. Fthenakis, G. Somatic cell counts in milk of Welsh-Mountain, Dorset-Horn and Chios ewes throughout lactation. *Small Rumin. Res.* **1996**, *20*, 155–162. [CrossRef]

39. Carpino, S.; Mallia, S.; La Terra, S.; Melilli, C.; Licitra, G.; Acree, T.; Barbano, D.M.; Van Soest, P. Composition and Aroma Compounds of Ragusano Cheese: Native Pasture and Total Mixed Rations. *J. Dairy Sci.* **2004**, *87*, 816–830. [CrossRef]

40. Segato, S.; Balzan, S.; Elia, C.A.; Lignitto, L.; Granata, A.; Magro, L.; Contiero, B.; Andrighetto, I.; Novelli, E. Effect of period of milk production and ripening on quality traits of Asiago cheese. *Ital. J. Anim. Sci.* **2007**, *6*, 469–471. [CrossRef]

41. Cozzi, G.; Ferlito, J.; Pasini, G.; Contiero, B.; Gottardo, F. Application of Near-Infrared Spectroscopy as an Alternative to Chemical and Color Analysis to Discriminate the Production Chains of Asiago d'Allevo Cheese. *J. Agric. Food Chem.* **2009**, *57*, 11449–11454. [CrossRef]

42. Cardinault, N.; Doreau, M.; Poncet, C.; Nozière, P. Digestion and absorption of carotenoids in sheep given fresh red clover. *Anim. Sci.* **2006**, *82*, 49–55. [CrossRef]

43. Palmquist, D.L. Milk Fat: Origin of Fatty Acids and Influence of Nutritional Factors Thereon. In *Advanced Dairy Chemistry Volume 2 Lipids*; Springer Science and Business Media LLC: Berlin/Heidelberg, Germany, 2007; pp. 43–92.

44. Bauman, D.E.; Griinari, J.M. Nutritional regulation of milk fat synthesis. *Annu. Rev. Nutr.* **2003**, *23*, 203–227. [CrossRef]

45. MacGibbon, A.; Taylor, M.W. Composition and Structure of Bovine Milk Lipids. In *Advanced Dairy Chemistry Volume 2 Lipids*; Springer Science and Business Media LLC: Berlin/Heidelberg, Germany, 2007; Volume 2, pp. 1–42.

46. Mele, M. Designing milk fat to improve healthfulness and functional properties of dairy products: From feeding strategies to a genetic approach. *Ital. J. Anim. Sci.* **2009**, *8*, 365–374. [CrossRef]

47. Dewhurst, R.J.; Shingfield, K.; Lee, M.; Scollan, N. Increasing the concentrations of beneficial polyunsaturated fatty acids in milk produced by dairy cows in high-forage systems. *Anim. Feed. Sci. Technol.* **2006**, *131*, 168–206. [CrossRef]

48. Trani, A.; Gambacorta, G.; Loizzo, P.; Cassone, A.; Faccia, M. Short communication: Chemical and sensory characteristics of Canestrato di Moliterno cheese manufactured in spring. *J. Dairy Sci.* **2016**, *99*, 6080–6085. [CrossRef] [PubMed]

49. Altomonte, I.; Conte, G.; Serra, A.; Mele, M.; Cannizzo, L.; Salari, F.; Martini, M. Nutritional characteristics and volatile components of sheep milk products during two grazing seasons. *Small Rumin. Res.* **2019**, *180*, 41–49. [CrossRef]

50. Caprioli, G.; Nzekoue, F.K.; Fiorini, D.; Scocco, P.; Trabalza-Marinucci, M.; Acuti, G.; Tardella, F.M.; Sagratini, G.; Catorci, A. The effects of feeding supplementation on the nutritional quality of milk and cheese from sheep grazing on dry pasture. *Int. J. Food Sci. Nutr.* **2019**, *71*, 50–62. [CrossRef] [PubMed]

51. Fusaro, I.; Giammarco, M.; Odintsov Vaintrub, M.; Chincarini, M.; Manetta, A.C.; Mammi, L.M.E.; Palmonari, A.; Formigoni, A.; Vignola, G. Effects of three different diets on the fatty acid profile and sensory properties of fresh Pecorino cheese "Primo Sale". *Asian-Australasian J. Anim. Sci.* **2020**. [CrossRef] [PubMed]

52. Gómez-Cortés, P.; Frutos, P.; Mantecon, A.R.; Juárez, M.; De La Fuente, M.A.; Hervás, G. Addition of Olive Oil to Dairy Ewe Diets: Effect on Milk Fatty Acid Profile and Animal Performance. *J. Dairy Sci.* **2008**, *91*, 3119–3127. [CrossRef]

53. Gómez-Cortés, P.; Frutos, P.; Mantecon, A.R.; Juárez, M.; De La Fuente, M.; Hervás, G. Milk Production, Conjugated Linoleic Acid Content, and In Vitro Ruminal Fermentation in Response to High Levels of Soybean Oil in Dairy Ewe Diet. *J. Dairy Sci.* **2008**, *91*, 1560–1569. [CrossRef]

54. Gómez-Cortés, P.; Bach, A.; Luna, P.; Juárez, M.; De La Fuente, M.A. Effects of extruded linseed supplementation on n-3 fatty acids and conjugated linoleic acid in milk and cheese from ewes. *J. Dairy Sci.* **2009**, *92*, 4122–4134. [CrossRef]

55. Gómez-Cortés, P.; De La Fuente, M.; Toral, P.G.; Frutos, P.; Juárez, M.; Hervás, G. Effects of different forage:concentrate ratios in dairy ewe diets supplemented with sunflower oil on animal performance and milk fatty acid profile. *J. Dairy Sci.* **2011**, *94*, 4578–4588. [CrossRef]

56. Toral, P.G.; Frutos, P.; Hervás, G.; Gómez-Cortés, P.; Juarez, M.; De La Fuente, M. Changes in milk fatty acid profile and animal performance in response to fish oil supplementation, alone or in combination with sunflower oil, in dairy ewes. *J. Dairy Sci.* **2010**, *93*, 1604–1615. [CrossRef] [PubMed]

57. Toral, P.G.; Gómez-Cortés, P.; Frutos, P.; de la Fuente, M.A.; Juárez, M.; Hervás, G. Milk production and fatty acid profile after three weeks of diet supplementation with sunflower oil and marine algae in dairy ewes. In *Challenging Strategies to Promote the Sheep and Goat Sector in the Current Global Context*; Ranilla, M.J., Carro, M.D., Ben Salem, H., Morand-Fehr, P., Eds.; Options Méditerranéennes: Série A. Séminaires Méditerranéens; n. 99; CIHEAM: Bari, Italy, 2011; pp. 337–342.

58. Vecchio, R.; Lombardi, A.; Cembalo, L.; Caracciolo, F.; Cicia, G.; Masucci, F.; Di Francia, A. Consumers' willingness to pay and drivers of motivation to consume omega-3 enriched mozzarella cheese. *Br. Food J.* **2016**, *118*, 2404–2419. [CrossRef]

59. Collomb, M.; Bütikofer, U.; Sieber, R.; Jeangros, B.; Bosset, J.-O. Correlation between fatty acids in cows' milk fat produced in the Lowlands, Mountains and Highlands of Switzerland and botanical composition of the fodder. *Int. Dairy J.* **2002**, *12*, 661–666. [CrossRef]

60. Falchero, L.; Lombardi, G.; Gorlier, A.; Lonati, M.; Odoardi, M.; Cavallero, A. Variation in fatty acid composition of milk and cheese from cows grazed on two alpine pastures. *Dairy Sci. Technol.* **2010**, *90*, 657–672. [CrossRef]

61. Kelsey, J.; Corl, B.; Collier, R.; Bauman, D. The Effect of Breed, Parity, and Stage of Lactation on Conjugated Linoleic Acid (CLA) in Milk Fat from Dairy Cows. *J. Dairy Sci.* **2003**, *86*, 2588–2597. [CrossRef]

62. Braghieri, A.; Piazzolla, N.; Romaniello, A.; Paladino, F.; Ricciardi, A.; Napolitano, F. Effect of adjuncts on sensory properties and consumer liking of Scamorza cheese. *J. Dairy Sci.* **2015**, *98*, 1479–1491. [CrossRef]

63. Khattab, A.R.; Guirguis, H.A.; Tawfik, S.M.; Farag, M.A. Cheese ripening: A review on modern technologies towards flavor enhancement, process acceleration and improved quality assessment. *Trends Food Sci. Technol.* **2019**, *88*, 343–360. [CrossRef]

64. Farruggia, A.; Pomiès, D.; Coppa, M.; Ferlay, A.; Verdier-Metz, I.; Le Morvan, A.; Bethier, A.; Pompanon, F.; Troquier, O.; Martin, B. Animal performances, pasture biodiversity and dairy product quality: How it works in contrasted mountain grazing systems. *Agric. Ecosyst. Environ.* **2014**, *185*, 231–244. [CrossRef]

65. Coppa, M.; Verdier-Metz, I.; Ferlay, A.; Pradel, P.; Didienne, R.; Farruggia, A.; Montel, M.-C.; Martin, B. Effect of different grazing systems on upland pastures compared with hay diet on cheese sensory properties evaluated at different ripening times. *Int. Dairy J.* **2011**, *21*, 815–822. [CrossRef]

66. Sacchi, R.; Marrazzo, A.; Masucci, F.; Di Francia, A.; Serrapica, F.; Genovese, A. Effects of Inclusion of Fresh Forage in the Diet for Lactating Buffaloes on Volatile Organic Compounds of Milk and Mozzarella Cheese. *Molecules* **2020**, *25*, 1332. [CrossRef]

67. McSweeney, P.L.H. Biochemistry of cheese ripening. *Int. J. Dairy Technol.* **2004**, *57*, 127–144. [CrossRef]

68. Lappalainen, R.; Kearney, J.; Gibney, M. A pan EU survey of consumer attitudes to food, nutrition and health: An overview. *Food Qual. Preference* **1998**, *9*, 467–478. [CrossRef]

Article

Feeding, Muscle and Packaging Effects on Meat Quality and Consumer Acceptability of Avileña-Negra Ibérica Beef

Marta Barahona [1,*], Mohamed Amine Hachemi [1], José Luis Olleta [1], María del Mar González [2] and María del Mar Campo [1]

[1] Department of Animal Production and Food Science, Instituto Agroalimentario (IA2), Universidad de Zaragoza-CITA, Miguel Servet 177, 50013 Zaragoza, Spain; hachemi.inmv@gmail.com (M.A.H.); olleta@unizar.es (J.L.O.); marimar@unizar.es (M.d.M.C.)
[2] Asociación Española de Raza Avileña-Negra Ibérica, Padre Tenaguillo 8, 05004 Ávila, Spain; consejoregulador@carnedeavila.org
* Correspondence: martabm@unizar.es

Received: 19 May 2020; Accepted: 26 June 2020; Published: 30 June 2020

Abstract: In order to achieve an attractive and differentiated product for the consumer and to optimize and to maximize profitability for the farmers within the EU Protected Geographical Indication "Carne de Ávila", 24 yearling males of Avileña-Negra Ibérica breed were used to evaluate the effect of 2 feeding systems, concentrate (CON) and maize silage (SIL), and 2 packaging systems, vacuum (VAC) and modified atmosphere (MAP), on fatty acid composition, proximate analysis, water holding capacity and consumer acceptability of meat in 2 muscles: *Longissimus thoracis* (LT) and *Semitendinosus* (ST). Animals fed with concentrate showed higher carcass weight. However, feeding did not affect the proximate analysis of the meat. The use of maize silage improved the amount of conjugated linoleic acid and *n*-3 polyunsaturated fatty acids (PUFA) and the relation *n*-6 PUFA/*n*-3 PUFA. In LT muscle, feeding influenced texture, samples from SIL being more tender. The VAC packaging showed higher cooking losses than MAP in both muscles. Aging increased tenderness and cooking losses but decreased thawing losses. LT samples from SIL feeding were better accepted by consumers and VAC packaging showed higher scores than MAP. We can conclude that the use of maize silage could be an alternative feeding for this type of animals improving some aspects of the quality of the meat.

Keywords: concentrate; silage; modified atmosphere; vacuum; texture; fatty acids; water holding capacity; consumer acceptability

---

1. Introduction

During the past decades, consumers have shown an increasing interest in food production including animal origin [1] at the same time that beef consumption has decreased. Besides the increasing numbers of vegetarians and vegans [2,3], several factors related to the quality of the product have motivated this negative tendency such as animal health [4], origin [5,6], genetic factors [7,8], feeding systems [9–12], conditioning and processing of the meat [13], among others.

The European Union includes within the label Protected Geographical Indication (PGI) those products that are reared in traditional or well-defined production systems, which in the end promotes a clear diversification of agricultural production and specific products [14]. In that sense, the Avileña-Negra Ibérica local beef breed (*Bos taurus*) is traditionally produced in the western region of Spain and since 1996 shows PGI label [15]. After weaning, calves are generally fattened in confinement and fed with cereal straw and concentrates, until they reach around 12–14 months of age with approximately 500 kg of live weight (550 kg for entire males and 450 kg for females), obtaining a meat that improves its quality with aging [16].

The producers of beef in several countries have had an important crisis of profitability in recent years due to the high price of some raw materials for feeding their animals and the high purchase cost of the calves. More extensive production systems, with pasture or forage, have lowered costs compared with diets based on concentrates. However, the agro-climatic conditions of an area or the farm structure do not always allow rearing efficiently the animals in those systems. Several studies have analysed carcass and meat quality of this local breed alone or compared with others [16–20]. The forages are cheaper than concentrates and allow an integration of the animal into sustainable rural environments; moreover, they are rich in natural antioxidants, plant pigments and *n*-3 polyunsaturated fatty acids if they are composed of pasture grasses and legumes [21]. However, feeding diets based on silage are a feasible option because they can easily be produced in irrigated areas located near the feedlots where grazing is not an option. Including maize silage in diets given as total mixed rations for fattening cattle has aroused much interest in recent years [22–25]. Maize silage lowers the cost of rations by increasing forage consumption without decreasing energy concentration, while the risk of low roughage supply due to improper processing in the mixer wagon can easily be avoided by adding small amounts of grass hay or cereal straw to the diet [26].

On the other hand, there is a gradual appearance of brands that want to guarantee the quality of the product: Quality and safety can be the key to the future. Consumers have evolved in their criteria with respect to food selection. The type of packaging used will influence in the decision to choose meat from a refrigerated display cabinet at purchase [27]. Packaging systems vary in the equipment required, the visual appeal to consumers, shelf life and effects on eating quality. There is a tendency to select packaged meat with modified atmospheres, because the high content of oxygen produces a red brilliant colour in the surface of the meat, which is generally considered desirable by consumers.

Thus, the objectives of this work were to study the effect of the production system (concentrate vs. maize silage) and packaged method (MAP and vacuum) on instrumental beef quality and acceptability of Avileña-Negra Ibérica breed within Carne de Ávila IGP label, in two muscles (*Longissimus thoracis* and *Semitendinosus*). In this sense, the knowledge of the relationships among production, meat quality, instrumental and sensory quality could help improving the final product, as well as understanding consumer acceptability.

2. Materials and Methods

2.1. Experimental Design and Animal Management

This study was performed with animals reared in a farm under commercial practices following national regulations in animal welfare (Council Directive 2008/119/EC), with animals being slaughtered at a commercial abattoir with captive bolt stunning in agreement with regulation 1099/2009 of the European Union about protection of animals at the time of killing.

The experimental treatments were conducted with yearlings from Avileña-Negra Ibérica breed in Spain. Twenty-four young bulls (250.7 kg ± 64.7 standard deviation initial live weight and 200.5 days old ± 44.9 standard deviation) were selected randomly from commercial indoor batches of 200 animals and controlled through the trial. Groups of four bulls were reared in separate pens for a total of 3 pens per group with 4.8 m^2 space allowance per animal. These were assigned to two groups: One group of 12 animals (CON) was fed with concentrates and cereal straw ad libitum, coincidentally with the most common husbandry conditions in the country for intensive rearing, and the other group of 12 animals (SIL) was fed with a mixture of 70% of maize silage and 30% of concentrate also ad libitum using a unifeed mixer wagon. The raw components of the CON group diets were barley, corn, dried waste from corn distillery, rapeseed flour, soybean flour, palm oil, calcium carbonate, palm fatty acids salt, sodium carbonate, salt and magnesium oxide. In the case of the SIL group, the ingredients used were similar with the exception of barley that was not included, and the rapeseed flour was replaced by shelled soybeans. Water was administrated ad libitum in both groups and days in trial were 276 in CON and 233 in SIL.

The composition of these diets per 100 g was CON: 379 kcal of calorific value, 8.03 g of moisture, 6.73 g of ash, 7.88 g of fat, 14.6 g of crude protein, 70.8 g of total carbohydrates and SIL: 205 kcal of calorific value, 50.7% moisture, 5.51 g of ash, 5.90 g of fat, 12.3 g of crude protein, 76.2 g of total carbohydrates. Fatty acid composition (g/100 g of total fat) is presented in Table 1.

Table 1. Fatty acids (g/100 g of total fat) of the diets (CON: concentrate; SIL: 70% maize silage + 30% concentrate).

	CON	SIL
C6:0	-	0.04
C8:0	-	0.02
C12:0	0.13	0.11
C14:0	0.44	0.27
C16:0	24.14	16.59
C16:1	0.14	0.13
C17:0	-	0.07
C17:1	-	0.03
C18:0	2.81	2.35
C18:1	31.91	29.27
C18:2	38.00	46.51
C18:3	1.43	1.79
C20:0	0.43	0.61
C20:1	0.29	0.23
C22:0	-	0.15

2.2. Slaughter and Meat Sampling

The target for slaughter was visual fatness for this breed under commercial practices, performed at the farm by an expert, reaching a minimum of 3.5 on a 5-point-scale [28]. After a finishing period of approximately of 250 days, the bulls with the average weight of 578.2 kg ± 36.4 standard deviation kg and 455.2 days old ± 41.6 standard deviation the animals were slaughtered in an EU licensed abattoir, and carcass conformation and fatness were obtained according to EU regulations [29]. Carcasses were chilled at 4 °C under commercial conditions. At 72 h post-mortem, *Longissimus thoracis* (LT) and *Semitendinosus* (ST) muscles were removed from the left side of each carcass and immediately vacuum packaged. They were stored and transported under refrigeration at 4 °C to the Veterinary Faculty of the University of Zaragoza.

At 7 days post-mortem, steaks from each animal were obtained for the analysis of fatty acids, texture, water holding capacity and consumer's test.

2.3. Proximate Composition and Fatty Acids

Steaks (1 cm thick) for chemical composition were removed between T6 and T7 from LT and from the distal side of ST. The samples were packaged under vacuum conditions and immediately frozen at −18 °C for posterior analysis. The composition of moisture [30], ash [31], lipid [32] and protein [33] was analysed according to official methods. For the fatty acid analysis, intramuscular fat was extracted in chloroform: methanol [34]. The methyl ester preparation included KOH in methanol, with C19:0 as an internal standard and was analysed by gas chromatography in HP 6890 equipped with an ionization flame and an automatic injection system (HP 7683) and fitted with a SP 2380 column (100 m × 0.25 mm × 0.20 μm) and oven temperature programming as follows: column temperature was set at 140 °C, then raised at rate of 3 °C/min from 140 °C to 158 °C, and 1 °C /min to 165 °C, kept for 10 min, raised at 5 °C /min up to 220 °C and kept constant for 50 min. Inlet temperature was kept at 230 °C and detector at 240 °C. A split mode injector with split ratio of 1/32 was applied; nitrogen as the carrier gas was used at a constant flow rate of 0.8 mL/min with an injected volume of 1 μL; methyl esters were identified using retention times of Sigma chemical Co; Standards [35]. Each sample was analysed in duplicate, and results were expressed as a percentage of total fatty acids.

2.4. Texture Analysis

Texture was measured at 7, 14 and 21 days of aging. Steaks with 7 days of aging from both muscles were vacuum packaged and immediately frozen at −18 °C. Half of the steaks for 14 and 21 days of aging were vacuum packaged and kept in the dark in the refrigerator to avoid oxidation. The other half were packaged with modified atmosphere (MAP) with 70% O_2:30% CO_2, remaining displayed at 4 °C and light (1200 luxes for 12 h/day) in a commercial cabinet with doors to promote oxidation, in conditions that are also commercially used. When reaching 14 or 21 days of aging, the steaks in MAP were repackaged in vacuum bags, frozen and kept at −18 °C until analysed.

Samples were thawed 24 h before the analysis at 4 °C in batches of 12 steaks selected randomly. Texture analysis was performed using two methods: Warner Bratzler Shear Force (WBSF) in cooked meat and compression in raw meat. In the analysis of WBSF, the steaks were cooked under vacuum in a water bath at 75 °C, until reaching an internal temperature of 70 °C, and then cooled in cold water. After cooling, samples were cut in the direction of the muscle fibres, and an average of six parallelepiped samples of 1 cm^2 cross section were obtained and assessed with an Instron Texturometer 4301 equipped with a WBSF cell. Samples were placed so that the muscle fibre direction was perpendicular to the load cell. Values obtained were the shear force (kg) and toughness (kg/cm^2), considered as the energy required cutting the sample at the point of maximum stress), collecting the average of a minimum of four replicates per each sample. Texture of raw meat was analysed using a modified compression device that hinders transversal elongation of the sample [36], also with an average of six parallelepipeds samples of 1 cm^2 cross section. Values were recorded at 20% of compression rate (C20; N/cm^2), which is related to the strength of the muscle fibre, and 80% of compression rate (C80; N/cm^2), which is related to the strength of the connective tissue.

2.5. Water Holding Capacity (WHC)

Steaks intended for texture analysis were used to measure the WHC as water losses during conservation. Water losses were calculated by the difference in weights measured before and after exposure of the steaks during 14 and 21 days of aging (7 and 14 days of conservation respectively) at vacuum and MAP. Water losses during thawing and during cooking were calculated from the differences in weights previously obtained and subsequently thawing and cooking steaks 7, 14 and 21 days of ripening MAP and vacuum packed. The WHC is calculated according to the equation: WHC (%) = [(initial weight − weight final)/initial weight] × 100.

2.6. Consumer Analysis

The samples for consumer test of both muscles (LT and ST) were packaged in MAP and vacuum and remained on display in a refrigerator at 4 °C in light (those packaged in MAP) and in the darkness (those packaged in vacuum) until reaching 14 days of aging. Then, all of them were vacuum packaged and frozen at −18 °C until the day of analysis.

Samples were thawed and kept at 4 °C and darkness 24 h before the analysis. For cooking, steaks were kept at room temperature for one hour before they were placed in aluminium foil, coded with three-digit number assigned randomly and cooked in an industrial double plate grill SAMMIC GRS-5, pre-heated at 200 °C, without adding any salt or spices, until reaching an internal temperature of 70 °C controlled by internal thermocouple (JENWAY, 2000. Staffordshire, UK).

From each cooked sample and after removing the external fat, 10 pieces of equal size (2 × 2 × 2 cm) were obtained. They were wrapped in aluminium foil previously labelled with a three-digit number and maintained at 50 °C until presented to the consumers, delaying less than 10 min.

A total of 120 consumers participated in the analysis. Each consumer received eight samples in different order at random inside each muscle, so that the first four samples were from the LT, followed by four from the ST. The acceptability of tenderness, flavour and overall appraisal on a scale of nine points were assessed from 1 (I dislike extremely) to 9 (I like extremely). The midpoint of the scale was removed to force a positive or negative decision by the consumer [37].

2.7. Statistical Analysis

All data were analysed using the SPSS 22.0 statistical package. The analysis of fatty acids and proximate analysis were analysed by the GLM procedure considering feeding and muscle as fixed effects. Texture measurements, water holding capacity and consumers test were divided by muscle and analysed with the GLM procedure considering feeding, packaging and aging as fixed effects and covariance by carcass weight. In the consumer test, consumer was considered a random effect and carcass weight as a covariate. The Duncan's multiple range test with a significance of $p < 0.05$ was used to assess differences between average values

3. Results and Discussion

3.1. Production Traits and Carcass Quality

Animals were introduced in the experimental phase with a similar age and a small difference in live weight ($p < 0.1$), which showed a tendency to be bigger in the maize silage group than in the one with concentrate (273.83 kg vs. 227.67 kg). However, slaughter age, slaughter live weight and carcass weight differed significantly between lots, with higher average values in CON group than in SIL group: 473.17 days, 595.67 kg and 341.80 kg vs. 437.17 days, 560.67 kg and 318.37 kg, respectively (Table 2). This was probably due to the chosen target for slaughter, which was visual fatness before slaughtering and was reflected in there being no differences in carcass conformation or fatness. These two characteristics are essential for assessing the final price of the carcass [29].

Table 2. Production traits and meat quality of Avileña-Negra Ibérica yearlings feeding with concentrate (CON) or 70% maize silage + 30% concentrate (SIL).

	CON	SIL	SEM	Significance
n	12	12		
Initial age d	196.8	204.2	9.36	0.696
Initial weight kg	227.7	273.8	12.58	0.080
Slaughter age d	473.2 a	437.2 b	7.79	0.031
Slaughter weight kg	595.7 a	560.7 b	6.62	0.015
Carcass Weight kg	341.8 a	318.4 b	3.82	0.006
ADG kg/d	1.34	1.26	0.04	0.298
Conformation	8.58 (R+)	7.83 (R)	0.206	0.083
Fatness	8.00 (3)	7.42(2+)	0.179	0.117
Carcass yield %	57.4	56.8	0.287	0.304

SEM: Standard error mean. a, b: different letters indicate significant differences in the mean values ($p < 0.05$).

No differences were found in average daily gain (ADG), similar in both groups. The same results were obtained by Casasús et al. (2012) [25] in a study comparing the use of unifeed with 80% maize silage and 20% concentrate and Avilés et al. (2015) [20] with a diet unifeed composed of the same proportion of concentrate, maize silage and wheat straw. On the contrary, other authors have found that the ADG of animals fed with concentrate was bigger than in those fed with a base of grass silage [38]. Furthermore, Steen et al. [39] found a higher ADG (38%) in animals receiving grass silage and concentrate that in the animals fed only with grass silage, because of the higher energy content of the concentrate. This might have also happened in our experiment; although ADG was not significantly different between treatments, CON animals were longer in the trial and reached a significantly heavier weight at slaughter (Table 2), with a diet higher in energy.

The ADG is similar to that found by Campo et al. [16] for this breed but much lower than that published by Piedrafita et al. [40] with 1.64 kg/d. The slaughter age in our study was a bit older than the slaughter age in the latter case, which was between 12 and 13 months, and this could explain these differences. When slaughtering animals older than 14 months, the growth in the last stage is based in fat deposition and, therefore, the weight increase slows down. However, in our study, there were not significant differences in carcass fat deposition (Table 1). The conformation tended to be slightly higher in animals fed with concentrates than in animals with maize silage, probably as a result of the higher carcass weight. Similar results observed Realini et al. [21] in a study comparing animals finished with grass with animals finished with silage. Moreover, Casasús et al. [25] observed better conformation in animals fed with concentrates, again due to the higher energy in the feeding.

3.2. Proximate Analysis

Diet did not have an effect on proximate composition (Table 3), but the type of muscle had a significant effect in the percentage of ashes. Previous studies with the same breed showed higher intramuscular fat in meat from animals fed under free-range conditions and supplemented with concentrate than in meat from those finished in confinement with cereal straw and concentrate [15]. The target of similar fatness at slaughter might be the responsible for this lack of differences in our study. The amount of intramuscular fat in beef from Avileña-Negra Ibérica breed was around 4%. This value would be classified as lean meat according to Food Advisory committee (1990) [41]. However, other authors found lower values in studies carried out with this same breed [16,40].

Table 3. Effect of feeding (concentrated (CON) and 70% maize silage + 30% concentrate (SIL)) and muscle (*Longissimus thoracis* (LT) and *Semitendinosus* (ST)) on proximate analysis of meat from Avileña-Negra Ibérica breed.

	FEEDING		MUSCLE			Significance		
	SIL	CON	LT	ST	SEM	FEED	MUS	FEED × MUS
n	24	24	24	24				
Moisture	72.5	73.0	72.5	72.9	0.207	0.401	0.499	0.833
Protein	21.9	22.2	22.0	22.1	0.208	0.409	0.802	0.802
Lipid	4.47	3.63	4.36	3.73	0.277	0.138	0.261	0.766
Ashes	1.16	1.18	1.10 b	1.24 a	0.014	0.710	<0.001	0.113

SEM: standard error mean; a, b: different letters indicate significant differences in the mean values ($p < 0.05$) within effects.

Some authors [42] have reported a linear relationship between fat and protein content. Acheson et al. [42] reported that ash values increase with fat as quality grade increases, and higher quality grade is positively correlated with higher intramuscular content. Although the ash content was significantly higher in ST than in LT (1.24% vs. 1.10%, respectively), we have not found this relationship with protein or fat content, since no statistical differences were found.

3.3. Fatty Acids

The fatty acid composition of the muscles and diets is presented in Table 4. LT showed higher percentage of saturated fatty acids than ST. These differences were mainly due to the amounts of the more common saturated fatty acids (SFA) presented in the meat, such as myristic acid (C14:0), palmitic acid (C16:0) and stearic acid (C18:0) at the cost of polyunsaturated fatty acids (PUFA). There were no differences in monounsaturated fatty acids (MUFA) between muscles. However, the percentages of *t*C18:1*n*-10+11 and C17:1 were significantly higher in LT than in ST. Oleic acid (C18:1-*n*-9) was similar in both muscles and C18:1*n*-11 was higher in ST than LT muscle.

Table 4. Effect of muscle *Longissimus thoracis* (LT) and *Semitendinosus* (ST) and feeding (concentrate (CON) and 70% maize silage + 30% concentrate (SIL)) on fatty acids composition (g/100 g of total fatty acids) of meat from Avileña-Negra Ibérica breed.

	LT	ST	CON	SIL	SEM	Muscle	Feed	Feed × Mus
n	24	24	24	24				
C10:0	0.03 a	0.02 b	0.03	0.03	0.002	0.001	0.680	0.037
C12:0	0.04 a	0.03 b	0.04	0.03	0.002	0.031	0.071	0.558
C13:0	0.01 a	0.00 b	0.01	0.01	0.000	0.001	0.042	0.719
C14:0	2.09 a	1.59 b	1.91	1.77	0.070	<0.001	0.226	0.230
C15:0	0.27	0.24	0.28 a	0.23 b	0.006	0.049	<0.001	0.353
C16:0	24.39 a	22.54 b	23.11	23.83	0.264	<0.001	0.109	0.062
C17:0	0.85 a	0.76 b	0.86 a	0.74 b	0.017	0.003	<0.001	0.626
C18:0	18.10 a	13.44 b	14.96 b	16.58 a	0.496	<0.001	0.026	0.455
C19:0	0.17	0.16	0.16	0.17	0.004	0.194	0.157	0.829
C20:0	0.12 a	0.09 b	0.11	0.11	0.004	<0.001	0.526	0.869
C21:0	0.28	0.32	0.27 b	0.33 a	0.012	0.106	0.013	0.135
C22:0	0.05 b	0.08 a	0.04 b	0.09 a	0.019	<0.001	<0.001	0.001
C14:1	0.25	0.27	0.31 a	0.21 b	0.001	0.508	0.015	0.518
C15:1	0.01	0.01	0.01	0.01	0.001	0.563	0.026	0.626
C16:1	2.02	2.15	2.36 a	1.82 b	0.084	0.387	0.001	0.515
C17:1	0.45	0.52	0.58 a	0.40 b	0.087	0.043	<0.001	0.885
tC18:1$n-10+11$	2.74 a	2.27 b	2.48	2.53	0.087	0.006	0.783	0.470
C18:1$n-9$	31.37	30.77	32.82 a	29.32 b	0.529	0.535	0.001	0.941
C18:1$n-11$	1.33 b	1.53 a	1.57 a	1.28 b	0.035	<0.001	<0.001	0.506
C18:1$n-13$	0.24	0.24	0.20 b	0.28 a	0.018	0.884	0.023	0.703
C20:1	0.13	0.15	0.14	0.14	0.005	0.140	0.531	0.201
C22:1$n-9$	0.04	0.04	0.06 a	0.02 b	0.008	0.597	0.025	0.048
tC18:2$n-6$	0.13 b	0.16 a	0.13 b	0.16 a	0.005	0.005	0.017	0.401
C18:2$n-6$	9.31 b	13.65 a	11.22	11.74	0.521	<0.001	0.544	0.293
C20:2$n-6$	0.06 b	0.10 a	0.07 b	0.09 a	0.004	<0.001	<0.001	0.004
C20:2$n-3$	0.07 b	0.14 a	0.10	0.12	0.008	<0.001	0.075	0.091
C22:2$n-6$	0.01 b	0.02 a	0.01 b	0.02 a	0.001	<0.001	<0.001	0.066
C18:3$n-6$	0.05 b	0.07 a	0.05 b	0.07 a	0.003	0.001	<0.001	0.020
C18:3$n-3$	0.25 b	0.34 a	0.25 b	0.34 a	0.013	<0.001	<0.001	0.025
C20:3$n-6$	0.33 b	0.60 a	0.37 b	0.56 a	0.033	<0.001	<0.001	0.014
C20:3$n-3$	0.07	0.11	0.04 b	0.14 a	0.012	0.041	<0.001	0.726
C20:4$n-6$	1.98 b	3.67 a	2.57	3.08	0.185	<0.001	0.064	0.175
C20:5$n-3$	0.12 b	0.27 a	0.17	0.22	0.018	<0.001	0.087	0.059
C22:5$n-3$	0.29 b	0.69 a	0.39 b	0.58 a	0.042	<0.001	0.001	0.027
C22:6$n-3$	0.02 b	0.06 a	0.03 b	0.05 a	0.003	<0.001	0.009	0.135
SFA	46.47 a	39.63 b	41.91 b	44.18 a	0.638	<0.001	0.003	0.055
MUFA	38.57	37.94	40.51 a	35.99 b	0.639	0.581	<0.001	0.927
PUFA	12.98 b	20.21 a	15.68	17.51	0.806	<0.001	0.137	0.186
$n-6$	11.88 b	18.26 a	14.41	15.73	0.729	<0.001	0.249	0.212
$n-3$	0.82 b	1.62 a	0.99 b	1.45 a	0.089	<0.001	<0.001	0.035
PUFA/SFA	0.28 b	0.52 a	0.38	0.42	0.024	<0.001	0.318	0.135
$n-6/n-3$	14.82 a	12.08 b	15.37 a	11.53 b	0.464	<0.001	<0.001	0.358

SEM: standard error mean; a, b: different letters indicate significant differences in the mean values within effects ($p < 0.05$). SFA: Saturated fatty acids; MUFA: Monounsaturated fatty acids; PUFA: Polyunsaturated fatty acids.

The effect of muscle was significant for all the PUFA, with higher concentration in ST than in LT. The linoleic acid was the predominant PUFA presented in the muscle and significantly higher in ST than LT (13.65% vs. 9.31%, respectively). PUFA are predominant in "red" muscles, such as ST, due to the higher content of phospholipids in them in relation to white muscles [43]. Although LT is not a strictly "white" muscle because of the mix of fibres that it has, it is different enough from ST in its fibre composition.

The effect of the diet on fatty acid composition was significant for the total SFA and MUFA. The use of maize silage produced meat with higher percentage of $n - 3$ PUFA. However, the percentage of $n - 6$ PUFA was not affected by the diet, although SIL showed high content of 18:2 $n - 6$ in the feed. The variability between animals are responsible for the no differences between the average values of intramuscular fat (4.5% vs. 3.6%). A higher saturation of meat from animals fed with SIL could be due to the higher incorporation of fat in the neutral lipids fraction because in ruminant muscle most PUFA are located in the phospholipid fraction [43]. Other authors have reported similar results. Warren et al. [44] observed that the use of ray-grass silage decreased the relationship PUFA/SFA, increasing the saturation. That increase was because of the reduction of MUFA, especially the percentage of C18:1 $n - 9$ that was lower in animals fed SIL than in CON animals (29.32% vs. 32.82%, respectively) as a reflection of the lower content in the SIL diet vs. CON. Casasús et al. [25] also found a decrease of ratio $n - 6/n - 3$ because of the higher proportion of the $n - 3$ PUFA in the meat of animals fed with maize silage. The higher proportion of $n - 3$ PUFA is associated to the composition of the silage, different to the concentrate, because the cereals and soya had higher content of $n - 6$ PUFA than silage. Some authors have showed that the use of maize silage could cause an increase of SFA and PUFA at the same time that decreases the MUFA [45,46]. As a result of the higher composition of $n - 3$ PUFA, the ratio $n - 6$ PUFA/$n - 3$ PUFA was lower ($p < 0.001$) in SIL than in CON animals (11.5 vs. 15.4). These values are typical in intensively fed animals that do not graze [43] but are far from the recommended level below 4 in the diet in terms of human health [47]. The ratio PUFA/SFA, which should be over 0.4 in a healthy diet [47], was reached by both groups without significant differences between them.

3.4. Texture Analysis and Water Holding Capacity

There was an effect of the feed on maximum load and toughness in LT (Table 5). Meat from CON group was tougher than meat from SIL group. The meat texture is very important for satisfying the parameters required by consumers, especially in beef where this organoleptic characteristic is valued primarily for consumption [48]. Even more, Hoving-Bolink et al. [49] found higher meat tenderness in animals reared with maize silage than in those fed with grass silage. Texture, considered as WBSF, has a high variability, which is attributable to factors related to animals, environment, pre-slaughter or breed [16], sex [50], age at slaughter [51], slaughter weight [52], feeding system, feeding level [53], compensatory growth [54], average daily gain, physical activity or confinement time [55] as well as technological factors, maturation, packaging, temperature and cooking techniques.

Another indicator associated with meat tenderness is the water losses, where meats having greater water loss usually have lower tenderness [56]. Still, the water holding capacity decreases with the increase in age of the animal [57]. Some authors have found differences between breeds [58], and the breeds with larger average daily gain presented lower WHC.

As it is shown in Table 4, in the muscle LT, thawing losses were affected by packaging and aging. Samples packaged in MAP and aged for 7 days showed the higher thawing losses. However, water losses were higher in samples from VAC and aged for 21 days. Cooking losses were affected by feeding in muscle ST (Table 6), where the samples from CON showed higher losses than samples from SIL feeding. Thawing and display losses were affected by packaging and aging; the same as occurred in muscle LT.

The increase in water losses during display (LT and ST) over time may be a consequence of the proteolysis that takes places during tenderization [59], which decreases the retention force of water by structural proteins. Display losses were greater in meat in vacuum than in MAP, and this could be due to the negative pressure the muscle receives during vacuum, which could facilitate the loss of part of the exudate.

The least thawing loss in the most exposed meat could be because some of the water was lost during display prior to be frozen. The losses were similar for all three aging times, with a trend of increased average losses in meat from animals fed with maize silage when compared to the concentrate fed. On the other hand, in the ST, the type of feeding did not affect any type of water losses analysed.

Table 5. Effect of feeding (concentrate (CON) vs. 70% maize silage + 30% concentrate (SIL)) and packaging (vacuum (VAC) and modified atmosphere (MAP)) and aging (7, 14 and 21 days) on texture parameters (maximum load and toughness) water holding capacity (WHC) of thawing, cooking and display of *Longissimus thoracis* muscle of meat from Avileña-Negra Ibérica breed and significance of effect of feeding (FEED), packaging (PACK) and aging (AGING).

| | Longissimus thoracis | | | | | | | | | | |
| | FEED | | PACKAGING | | AGING | | | | Significance | | |
	CON	SIL	MAP	VAC	7 d	14 d	21 d	SEM	FEED	PACK	AGING
n	72	72	36	36	48	48	48				
Max load (Kg)	5.01 a	4.20 b	4.67	4.55	5.13 a	4.44 b	4.24 b	0.103	<0.001	0.537	<0.001
Toughness (Kg/cm^2)	1.87 a	1.69 b	1.83	1.74	1.80	1.77	1.78	0.030	<0.001	0.119	0.887
WHC thawing (%)	4.14	3.99	4.37 a	3.76 b	4.82 a	3.54 b	3.83 b	0.139	0.188	0.014	<0.001
WHC cooking (%)	27.89	26.95	27.07	27.77	27.86	26.50	27.90	0.366	0.279	0.290	0.150
WHC display (%)	4.07	4.48	3.77 b	4.79 a	-	3.95 b	4.61 a	0.163	0.510	0.001	0.032

Means with different letters are significantly different within effect ($p < 0.05$). SEM: standard error mean. Interactions not significant.

Table 6. Effect of feeding (concentrate (CON) vs. 70% maize silage + 30% concentrate (SIL)) and packaging (vacuum (VAC) and modified atmosphere (MAP)) and aging (7, 14 and 21 days) on texture parameters (maximum load and toughness), water holding capacity (WHC) of thawing, cooking and display of *Longissimus thoracis* muscle of meat from Avileña-Negra Ibérica breed and significance of effect of feeding (FEED), packaging (PACK) and aging (AGING).

| | Semitendinosus | | | | | | | | | | |
| | FEED | | PACKAGING | | AGING | | | | Significance | | |
	CON	SIL	MAP	VAC	7 d	14 d	21 d	SEM	FEED	PACK	AGING
n	72	72	36	36	48	48	48				
Max load (Kg)	4.22	4.07	4.28 a	4.01 b	4.34 a	4.12 a,b	3.98 b	0.057	0.291	0.019	0.031
Toughness (Kg/cm^2)	2.02	2.03	2.08	1.96	1.98	2.06	2.02	0.032	0.244	0.048	0.609
WHC thawing (%)	4.27	4.30	4.60 a	3.96 b	4.86 a	4.10 b	3.89 b	0.131	0.823	0.013	0.006
WHC cooking (%)	27.93 a	26.32 b	26.52	27.74	27.90	26.40	27.08	0.241	0.023	0.116	0.284
WHC display (%)	6.41	6.16	5.85 b	6.71 a	-	5.48 b	7.08 a	0.193	0.151	0.016	<0.001

Means with different letters are significantly different within effect ($p < 0.05$). SEM: standard error mean. Interactions not significant.

3.5. Consumer Analysis

Consumers' scores are presented in Tables 7 and 8. The results were separated by muscle. In the case of LT, the effect of the feed was significant for global and tenderness acceptability, being samples from SIL better accepted than CON. For flavour acceptability, feeding had no effect on consumer's perception. However, those differences were not appreciated in the muscle ST, where feeding had no effect on consumer acceptability. Other authors had also reported this lack of effect of the feeding [15,20,60].

Table 7. Effect of feeding (concentrate (CON) vs. 70% maize silage + 30% concentrate (SIL)) and packaging (vacuum (VAC) and modified atmosphere (MAP)) on global, tenderness and flavour acceptability of *Longissimus thoracis* (LT) muscle of meat from Avileña-Negra Ibérica breed.

LT	FEED		PACKAGING				Significance	
	CON	SIL	MAP	VAC	SEM	FEED	PACK	FEED × PACK
n	240	240	240	240				
GLOBAL	6.21	6.50	5.94 b	6.77 a	0.079	0.044	<0.001	0.930
TENDER	5.95 b	6.44 a	5.67 b	6.72 a	0.091	0.004	<0.001	0.459
FLAVOUR	6.47	6.58	6.12 b	6.91 a	0.077	0.326	<0.001	0.619

Scale of nine points: 1 = dislike extremely; 9 = like extremely. Means with different letters (a, b) are significantly different within effect ($p < 0.05$).

Table 8. Effect of feeding (concentrated (CON) vs. 70% maize silage + 30% concentrate (SIL)) and packaging (vacuum (VAC) and modified atmosphere (MAP)) on global, tenderness and flavour acceptability of *Semitendinosus (ST)* muscle of meat from Avileña-Negra Ibérica breed.

ST	FEED		PACKAGING				Significance	
	CON	SIL	MAP	VAC	SEM	FEED	PACK	FEED × PACK
n	240	240	240	240				
GLOBAL	5.64	5.54	5.01 b	6.16 a	0.099	0.329	<0.001	0.610
TENDER	5.61	5.57	5.06 b	6.12 a	0.103	0.131	<0.001	0.697
FLAVOUR	5.59	5.54	4.99 b	6.13 a	0.099	0.609	<0.001	0.100

Scale of nine points: 1 = dislike extremely; 9 = like extremely. Means with different letters (a, b) are significantly different within effect ($p < 0.05$).

In addition, greater acceptability notes of tenderness and flavour for LT when compared to the ST were observed. Moreover, the samples packaged in vacuum had higher global notes when comparing with the samples packaged in the MAP. MAP is rich in oxygen to improve and keep an attractive colour during display, but it favours oxidation. Oxidation provokes more rancid notes during conservation [61], and this can reduce flavour acceptability after consumption.

Feeding affected the acceptability of LT, where consumers preferred meat from SIL group. However, those differences only appeared for global acceptability and tenderness but not flavour. Tenderness is one of the sensory quality parameters most affected by aging since it increases tenderness due to proteolysis, and most consumers prefer tender meat [16]. On the other hand, sometimes meat packaged in MAP has been found less tender than vacuum packaged meat [62] similar to the findings of this study. This effect in high oxygen packaging could be explained by the decrease in the proteolysis of myofibrils, because of the decrease in the μ calpain activity [63]. Consumers found significant differences for packaging. In both muscles, the scores for the three attributes were higher than 6 in samples from vacuum packaging in comparison with MAP packaging, especially in muscle LT, in which the scores were higher than 6.5 in a 9 points scale.

4. Conclusions

The use of maize silage as an alternative to conventional concentrate during the confinement period did not alter the productive parameters in terms of average daily gain (ADG), carcass percentage and fatness score. Maize silage would be recommended for improving tenderness of Avileña-Negra ibérica beef, especially in the case of muscles with higher commercial category such as LT. However, aging of the meat it is still one of the best ways to improve its tenderness.

Regarding the sensory quality, vacuum aging improves overall, tenderness and flavour acceptability; moreover, the LT muscle shows a greater global and flavour acceptability when compared to the ST. Therefore, confinement beef with maize silage may be an alternative that meets the quality requirements of the market and of consumers of beef.

Author Contributions: Conceptualization, M.d.M.C. and J.L.O.; methodology M.A.H. and M.B.; formal analysis, M.A.H. and M.B.; investigation, M.d.M.C. and J.L.O.; resources, M.d.M.C.; data curation, M.d.M.C. and J.L.O.; writing—original draft preparation, M.B.; writing—review and editing, M.d.M.C. and J.L.O.; visualization, M.B.; supervision, M.d.M.C and J.L.O.; project administration, M.d.M.G.; funding acquisition, J.L.O. All authors have read and agreed to the published version of the manuscript.

Funding: This research was funded by Ministerio de Agricultura, Alimentación y Medio Ambiente, through the Funding Program 23.12.412C.770.02 (BOE 23 June 2014).

Acknowledgments: The authors would like to thank the INIA for funding the Project 20130020000829. They also thank the Carlos Sañudo and Animal Production personnel of the Faculty of Veterinary of Zaragoza, personnel of Carne de Ávila PGI and the consumers who participated in the trial.

Conflicts of Interest: The authors declare no conflict of interest.

References

1. Mørkbak, M.R.; Nordström, J. The impact of information on consumer preferences for different animal food production methods. *J. Consum. Policy* **2009**, *32*, 313–331. [CrossRef]
2. Clark, M.; Tilman, D. Comparative analysis of environmental impacts of agricultural production systems, agricultural input efficiency, and food choice. *Environ. Res. Lett.* **2017**, *12*, 064016. [CrossRef]
3. Dubois, G.; Sovacool, B.; Aall, C.; Nilsson, M.; Barbier, C.; Herrmann, A.; Brùyere, S.; Andersson, C.; Skold, B.; Nadaud, F.; et al. It starts at home? Climate policies targeting household consumption and behavioral decisions are key to low-carbon futures. *Energy Res. Soc. Sci.* **2019**, *52*, 1297. [CrossRef]
4. Wezemael, L.V.; Verbeke, W.; Kügler, J.O.; Barcellos, M.D.; Grunert, K.G. European consumers and beef safety: Perceptions, expectations and uncertainty reduction strategies. *Food Control* **2010**, *21*, 835–844. [CrossRef]
5. Sañudo, C.; Macie, E.S.; Olleta, J.L.; Villarroel, M.; Panea, B.; Albertí, P. The effects of slaughter weight, breed type and ageing time on beef meat quality using two different texture devices. *Meat Sci.* **2004**, *66*, 925–932. [CrossRef]
6. Campo, M.M.; Brito, G.; Soares de Lima, J.M.; Vaz Martins, D.; Sañudo, C.; San Julián, R.; Hernández, P. Effects of feeding strategies including different proportion of pasture and concentrate, on carcass and meat quality traits in Uruguayan steers. *Meat Sci.* **2008**, *80*, 753–760. [CrossRef]
7. Campion, B.; Keane, M.G.; Kenny, D.A.; Berry, D.P. Evaluation of estimated genetic merit for carcass weight in beef cattle: Live weights, feed intake, body measurements, skeletal and muscular scores, and carcass characteristics. *Livest. Sci.* **2009**, *126*, 87–99. [CrossRef]
8. Calvo, J.H.; Iguácel, L.P.; Kirinus, J.K.; Serrano, M.; Ripoll, G.; Casusús, I.; Joy, M.; Pérez-Velasco, L.; Santo, P.; Albertí, P.; et al. A new single nucleotide polymorphism in the calpastatin (CAST) gene associated with beef tenderness. *Meat Sci.* **2014**, *96*, 775–782. [CrossRef]
9. Muir, P.D.; Deaker, J.M.; Bown, M.D. Effects of forage- and grain-based feeding systems on beef quality: A review. *N. Z. J. Agric. Res.* **1998**, *41*, 623–635. [CrossRef]
10. Franco, D.; Bispo, E.; González, L.; Vázquez, J.A.; Moreno, T. Effect of finishing and ageing time on quality attributes of loin from the meat of Holstein–Fresian cull cows. *Meat Sci.* **2009**, *83*, 484–491. [CrossRef]
11. De la Fuente, J.; Díaz, M.T.; Alvarez, I.; Oliver, M.A.; Font I Furnols, M.; Sañudo, C.; Campo, M.M.; Montossi, F.; Nute, G.R.; Cañeque, V. Fatty acid and vitamin E composition of intramuscular fat in cattle reared in different production systems. *Meat Sci.* **2009**, *82*, 331–337. [CrossRef]

12. Morales, R.; Parga, J.; Subiabre, I.; Realini, C.E. Finishing strategies for steers based on pasture or silage plus grain and time on feed and their effects on beef quality. *Int. J. Agric. Nat. Resour.* **2015**, *42*, 5–18. [CrossRef]

13. Murphy, K.M.; O'Grady, M.N.; Kerry, J.P. Effect of varying the gas headspace to meat ratio on the quality and shelf-life of beef steaks packaged in high oxygen modified atmosphere packs. *Meat Sci.* **2013**, *94*, 447–454. [CrossRef]

14. Serra, X.; Gil, M.; Gispert, M.; Guerrero, L.; Oliver, M.A.; Sañudo, C.; Campo, M.M.; Panea, B.; Olleta, J.L.; Quintanilla, R.; et al. Characterisation of young bulls of the Bruna dels Pirineus cattle breed (selected from old Brown Swiss) in relation to carcass, meat quality and biochemical traits. *Meat Sci.* **2004**, *66*, 425–436. [CrossRef]

15. Daza, A.; Rey, A.I.; Lopez-Carrasco, C.; Lopez-Bote, C.J. Effect of gender on growth performance, carcass characteristics and meat and fat quality of calves of Avileña-Negra Ibérica breed fattened under free-range conditions. *Span. J. Agric. Res.* **2014**, *12*, 683–693. [CrossRef]

16. Campo, M.M.; Sañudo, C.; Panea, B.; Alberti, P.; Santolaria, P. Breed type and ageing time effects on sensory characteristics of beef strip loin steaks. *Meat Sci.* **1999**, *51*, 383–390. [CrossRef]

17. Campo, M.M.; Santolaria, P.; Sañudo, C.; Lepetit, J.; Olleta, J.L.; Panea, B.; Albertí, P. Assessment of breed type and ageing time effects on beef meat quality using two different texture devices. *Meat Sci.* **2000**, *55*, 371–378. [CrossRef]

18. Díez, J.; Albertí, P.; Ripoll, G.; Lahoz, F.; Fernández, I.; Olleta, J.L.; Panea, B.; Sañudo, C.; Bahamonde, A.; Goyache, F. Using machine learning procedures to ascertain the influence of beef carcass profiles on carcass conformation scores. *Meat Sci.* **2006**, *73*, 109–115. [CrossRef]

19. Christensen, M.; Ertbjerg, P.; Failla, S.; Sañudo, C.; Richardson, R.I.; Nute, G.R.; Olleta, J.L.; Panea, B.; Albertí, P.; Juárez, M.; et al. Relationship between collagen characteristics, lipid content and raw and cooked texture of meat from young bulls of fifteen European breeds. *Meat Sci.* **2011**, *87*, 61–65. [CrossRef]

20. Avilés, C.; Martínez, A.L.; Domenech, V.; Peña, F. Effect of feeding system and breed on growth performance, and carcass and meat quality traits in two continental beef breeds. *Meat Sci.* **2015**, *107*, 94–103. [CrossRef]

21. Realini, C.E.; Duckett, S.K.; Brito, G.W.; Dalla Rizza, M.; De Mattos, D. Effect of pasture vs. concentrate feeding with or without antioxidants on carcass characteristics, fatty acid composition, and quality of Uruguayan beef. *Meat Sci.* **2004**, *66*, 567–577. [CrossRef]

22. Cooke, D.W.I.; Monahan, F.J.; Brophy, P.O.; Boland, M.P. Comparison of concentrates or concentrates plus forages in a total mixed ration or discrete ingredient format: Effects on beef production parameters and on beef composition, color, texture and fatty acid profile. *Ir. J. Agric. Food Res.* **2004**, *43*, 201–216.

23. Keady, T.W.; Lively, F.O.; Kilpatrick, D.J.; Moss, B.W. Effects of replacing grass silage with either maize or whole-crop wheat silages on the performance and meat quality of breed cattle offered two levels of concentrates. *Animal* **2007**, *1*, 613–623. [CrossRef]

24. Mazzenga, A.; Gianesella, M.; Brscic, M.; Cozzi, G. Feeding behaviour, diet digestibility, rumen fluid and metabolic parameters of beef cattle fed total mixed rations with a stepped substitution of wheat straw with maize silage. *Livest. Sci.* **2009**, *122*, 16–23. [CrossRef]

25. Casasús, I.; Ripoll, G.; Albertí, P. Use of maize silage in beef heifers fattening diets: Effects on performance, carcass and meat quality. *ITEA* **2012**, *108*, 191–206, In Spanish, with English abstract.

26. Cozzi, G.; Brscic, M.; Boukha, A.; Da Ronch, F.; Tenti, S.; Gottardo, F. Comparison of two feeding finishing treatments on production and quality of organic beef. *Ital. J. Anim. Sci.* **2010**, *9*, 404–409. [CrossRef]

27. McMillin, K.W. Advancements in meat packaging. *Meat Sci.* **2017**, *132*, 153–162. [CrossRef]

28. Edmonson, A.J.; Lean, I.J.; Weaver, L.D.; Farvert, T.; Webster, G. A body condition scoring chart for Holstein dairy cows. *J. Dairy Sci.* **1989**, *72*, 68–78. [CrossRef]

29. REGULATION (EU). *No 1308/2013 of the European Parliament and of the Council of 17 December 2013 Establishing a Common Organisation of the Markets in Agricultural Products and Repealing Council Regulations*; European Parliament and of the Council: Brussels, Belgium, 2013; L347; pp. 671–874.

30. ISO. *Meat and Meat Products—Determination of Moisture Content*; ISO 1442:1997; International Organization for Standardization: Geneva, Switzerland, 1997.

31. ISO. *Meat and Meat Products—Determination of Total Ash*; ISO 936:1998; International Organization for Standardization: Geneva, Switzerland, 1998.

32. ISO. *Meat and Meat Products—Determination of Total Fat Content*; ISO 1443:1973; International Organization for Standardization: Geneva, Switzerland, 1973.

33. ISO. *Meat and Meat Products—Determination of Nitrogen Content*; ISO 937:1978; International Organization for Standardization: Geneva, Switzerland, 1978.

34. Bligh, E.G.; Dyer, W.J. A rapid method of total lipid extraction and purification. *Can. J. Biochem.* **1959**, *37*, 911–917. [CrossRef] [PubMed]

35. Carrilho, M.C.; López, M.; Campo, M.M. Effect of the fattening diet on the development of the fatty acid profile in rabbits from weaning. *Meat Sci.* **2009**, *83*, 88–95. [CrossRef]

36. Lepetit, J.; Culioli, J. Mechanical-properties of meat. *Meat Sci.* **1994**, *36*, 203–237. [CrossRef]

37. Font i Furnols, M.; San Julián, R.; Guerrero, L.; Sañudo, C.; Campo, M.M.; Olleta, J.L.; Oliver, M.A.; Cañeque, V.; Álvarez, I.; Díaz, M.T.; et al. Acceptability of lamb meat from diferent producing systems and ageing time to German, Spanish and British consumers. *Meat Sci.* **2006**, *72*, 545–554. [CrossRef]

38. Caplis, J.; Keane, M.G.; Moloney, A.P.; O'Mara, F.P. Effects of supplementary concentrate level with grass silage, and separate or total mixed ration feeding, on performance and carcass traits of finishing steers. *Ir. J. Agric. Food Res.* **2005**, *44*, 27–43.

39. Steen, R.W.J.; Kilpatrick, D.J. The effects of the ratio of grass silage to concentrates in the diet and restricted dry matter intake on the performance and carcass composition of beef cattle. *Livest. Prod. Sci.* **2000**, *62*, 181–192. [CrossRef]

40. Piedrafita, J.; Quintanilla, R.; Sañudo, C.; Olleta, J.L.; Campo, M.M.; Panea, B.; Renand, G.; Turin, F.; Jabet, S.; Osoro, K.; et al. Carcass quality of ten beef cattle breeds of the south-west of Europe. *Livest. Prod. Sci.* **2003**, *82*, 1–13. [CrossRef]

41. Food Advisory Committee. *Report on Review of Food Labelling and Dyertising, 1990*; Her Majesty's Stationary Office: London, UK, 1990.

42. Acheson, R.J.; Woerner, D.R.; Martin, J.N.; Belk, K.E.; Engle, T.E.; Brown, T.R.; McNeill, S.H. Nutrient database improvement project: Separable components and proximate composition of raw and cooked retail cuts from the beef loin and round. *Meat Sci.* **2015**, *110*, 236–244. [CrossRef]

43. Wood, J.D.; Richardson, R.I.; Nute, G.R.; Fisher, A.V.; Campo, M.M.; Kasapidou, E.; Sheard, P.R.; Enser, M. Effects of fatty acids on meat quality: A review. *Meat Sci.* **2003**, *66*, 21–32. [CrossRef]

44. Warren, H.E.; Scollan, N.D.; Nute, G.R.; Hughes, S.I.; Wood, J.D.; Richardson, R.I. Effects of breed and a concentrate or grass silage diet on beef quality in cattle of 3 ages. II: Meat stability and flavour. *Meat Sci.* **2008**, *78*, 270–278. [CrossRef]

45. Łozicki, A.; Dymnicka, M.; Arkuszewska, E.; Pustkowiak, H. Effect of pasture or maize silage feeding on the nutritional value of beef. *Ann. Anim. Sci.* **2012**, *12*, 81–93. [CrossRef]

46. Moloney, A.P.; Mooney, M.T.; Kerry, J.P.; Stanton, C.; O'Kiely, P. Colour of fat, and colour, fatty acid composition and sensory characteristics of muscle from heifers offered alternative forages to grass silage in a finishing ration. *Meat Sci.* **2013**, *95*, 608–615. [CrossRef]

47. Department of Health. Nutritional Aspects of the Cardiovascular Disease. In *Report of Health and Social Subjects*; Her Majesty's Stationary Office: London, UK, 1994; Volume 46.

48. Koohmaraie, M.; Geesink, G.H. Contribution of post-mortem muscle biochemistry to the delivery of consistent meat quality with particular focus on the calpain system. *Meat Sci.* **2006**, *74*, 34–43. [CrossRef] [PubMed]

49. Hoving-Bolink, A.H.; Hanekamp, W.J.A.; Walstra, P. Effects of diet on carcass, meat and eating quality of once-bred Piemontese x Friesian heifers. *Livest. Prod. Sci.* **1999**, *57*, 267–272. [CrossRef]

50. Lucero-Borja, J.; Pouzo, L.B.; de la Torre, M.S.; Langman, L.; Carduza, F.; Corva, P.M.; Santini, F.J.; Pavan, E. Slaughter weight, sex and age effects on beef shear force and tenderness. *Livest. Sci.* **2014**, *163*, 140–149. [CrossRef]

51. Velik, M.; Steinwidder, A.; Frickh, J.J.; Ibi, G.; Kolbe-Römer, A. Effect of ration, sex and breed on carcass performance and meat quality of cattle from suckler cow systems. *Zuchtungskunde* **2008**, *80*, 378–388.

52. Lundesjö Ahnström, M.; Hessle, A.; Johansson, L.; Hunt, M.C.; Lundström, K. Influence of slaughter age and carcass suspension on meat quality in Angus heifers. *Anim. Sci. J.* **2012**, *6*, 1554–1562. [CrossRef]

53. Rezende, P.L.P.; Restle, J.; Fernandes, J.J.R.; Neto, M.D.F.; Prado, C.S.; Pereira, M.L.R. Características da carcaça e da carne de novilhos mestiços submetidos a diferentes estratégias nutricionais na recria e terminação. *Ciência Rural* **2012**, *4*, 875–881. [CrossRef]

54. Hansen, S.; Therkildsen, M.; Byrne, D.V. Effects of a compensatory growth strategy on sensory and physical properties of meat from young bulls. *Meat Sci.* **2006**, *74*, 628–643. [CrossRef]

55. Dashdorj, D.; Oliveros, M.C.R.; Hwang, I.H. Meat quality traits of longissimus muscle of Hanwoo steers as a function of interaction between slaughter endpoint and chiller ageing. *Korean J. Food Sci. Anim. Resour.* **2012**, *32*, 414–427. [CrossRef]

56. Dransfield, E.; Francombe, M.A.; Whelehan, O.P. Relationships between sensory attributes in cooked meat. *J. Text. Stud.* **1984**, *15*, 337–356. [CrossRef]

57. Wissmer-Pedersen, J. Química de los tejidos animales. Parte 5. Agua. In *Ciencia de la Carne y de los Productos Cárnicos*; Price, J.F., Schweigert, B.S., Eds.; Editorial Acribia: Zaragoza, Spain, 1994.

58. Albertí, P.; Sañudo, C.; Santolaria, P.; Negueruela, I.; Olleta, J.L.; Mamaqui, E.; Campo, M.M.; Alvarez, F.S. Calidad de la carne de terneros de raza Parda Alpina y Pirenaica cebados con pienso rico en gluten feed y mandioca. *ITEA* **1995**, *16*, 630–632.

59. Hughes, J.M.; Oiseth, S.K.; Purslow, P.P.; Warner, R.D. A structural approach to understanding the interactions between colour, water-holding capacity and tenderness. *Meat Sci.* **2014**, *98*, 520–532. [CrossRef] [PubMed]

60. Albertí, P.; Sañudo, C. Características de la canal y calidad de la carne de terneros cebados con dietas forrajeras y suplementados con distinta cantidad de pienso. *ITEA* **1988**, *19*, 3–14.

61. Campo, M.M.; Nute, G.R.; Hughes, S.I.; Enser, M.; Wood, J.D.; Richardson, I. Flavour perception of oxidation in beef. *Meat Sci.* **2006**, *72*, 301–311. [CrossRef]

62. Lagerstedt, A.; Lundström, K.; Lindahl, G. Influence of vacuum or high-oxygen modified atmosphere packaging on quality of beef M. longissimus dorsi steaks after different ageing times. *Meat Sci.* **2011**, *87*, 101–106. [CrossRef]

63. Rowe, L.J.; Maddock, K.R.; Lonergan, S.M.; Huff-Lonergan, E. Oxidative environments decrease tenderization of beef steaks through inactivation of μcalpain. *J. Anim. Sci.* **2004**, *82*, 3254–3266. [CrossRef] [PubMed]

Article

Composition, Mineral and Fatty Acid Profiles of Milk from Goats Fed with Different Proportions of Broccoli and Artichoke Plant By-Products

Paula Monllor [1], **Gema Romero** [1], **Alberto S. Atzori** [2], **Carlos A. Sandoval-Castro** [3], **Armín J. Ayala-Burgos** [3], **Amparo Roca** [1], **Esther Sendra** [1] and **José Ramón Díaz** [1,*]

[1] Departamento de Tecnología Agroalimentaria, Universidad Miguel Hernández de Elche, 03312 Alicante, Spain; pmonllor@umh.es (P.M.); gemaromero@umh.es (G.R.); aroca@umh.es (A.R.); esther.sendra@umh.es (E.S.)

[2] Dipartimento di Agraria, Università degli Studi di Sassari, 07100 Sassari, Italy; asatzori@uniss.it

[3] Facultad de Medicina Veterinaria y Zootecnia, Universidad Autónoma de Yucatán, Mérida 97100, Mexico; carlos.sandoval@correo.uady.mx (C.A.S.-C.); aayala@correo.uady.mx (A.J.A.-B.)

* Correspondence: jr.diaz@umh.es; Tel.: +34-966-749-707

Received: 28 April 2020; Accepted: 28 May 2020; Published: 1 June 2020

Abstract: In the Mediterranean region, artichoke and broccoli are major crops with a high amount of by-products that can be used as alternative feedstuffs for ruminants, lowering feed costs and enhancing milk sustainability while reducing the environmental impact of dairy production. However, nutritional quality of milk needs to be assured under these production conditions and an optimal inclusion ratio of silages should be determined. This work aimed to evaluate the effect of three inclusion levels (25%, 40%, and 60%) of these silages (artichoke plant, AP, and broccoli by-product, BB) in goat diets on milk yield, composition, and mineral and fatty profiles. Treatments with 60% inclusion of AP and BB presented the lowest milk yield. No differences were found on the milk mineral profile. Inclusion of AP in the animals' diet improved the milk lipid profile from the point of view of human health (AI, TI) compared to BB due to a lower saturated fatty acid content (C12:0, C14:0, and C16:0) and a higher concentration of polyunsaturated fatty acids (PUFA), especially vaccenic acid (C18:1 trans11) and rumenic acid (CLA cis9, trans11), without any differences with the control treatment.

Keywords: fatty acid profile; mineral profile; CLA; milk yield; circular economy

---

1. Introduction

Regarding milk consumption worldwide, cow's milk occupies first place, followed by buffalo and thirdly, that of goat [1], which continues to increase [2] due to its high level of calcium, phosphorus, and animal protein. In addition, goat milk has been classified as a substitute for cow's milk in those people who suffer from some type of allergy to this food [3]. Goat's milk is a source of nutrients in the human diet due to its content of Se and polyunsaturated fatty acids (PUFA), such as vaccenic and rumenic acid or CLA [4,5], which can influence the prevention of certain types of cancers and cardiovascular diseases [6,7]. The literature contains many studies of how diet affects the performance and quality of ruminant milk. Hilali et al. [8] and Cappucci et al. [9] found that the inclusion of agro-industrial and olive by-products in ewes' diets enhanced milk fatty acid profile, with no effects on performance and milk macro-composition. On the other hand, Schulz et al. [10] observed changes in milk fatty acid profile in cows fed with red clover silage in comparison with maize silage. Finally, Monllor et al. [11] showed slight differences in fat and protein levels of milk from goats fed with artichoke by-products and an increase of Selenium and polyunsurared fatty acid contents.

The inclusion of agricultural by-products in ruminant diets does not have to affect the sensory quality of dairy products. Such is the case in Caputo et al. Ref. [12], who did not observe differences in the aromatic profile of milk and dairy products from cows fed with destoned olive cake.

It is necessary to enhance the sustainability of milk production and reduce the impact of animal feeding. The use of local resources, especially if recovered from by-products, may significantly enhance milk sustainability. Artichoke (*Cynara scolymus* L.) and broccoli (*Brassica oleracea* var. Italica) crops generate large quantities of by-products. According to Food and Agriculture Organization of the United Nations (FAO) [13], 1,505,328 t of artichoke and 25,984,758 t of broccoli were harvested worldwide in 2017. The artichoke plant is a waste, mainly formed of stems and leaves, and some unharvested inflorescences are left in the field after harvest of inflorescences for human consumption. This by-product has traditionally been used by grazing small ruminants or collected and brought to dairy farms [14]. The yield of green fodder in this crop is 11.1 t/ha [15], which, taking into account FAO's cultivated area data [13] (2017) worldwide (122,390 ha), would result in an annual production of more than 1,300,000 t of available artichoke plant. According to Ros et al. [16], 29.5% of harvested broccoli is composed of stems and inflorescences that are not suitable for human consumption. Broccoli by-product is considered, from the point of view of animal feed, more as a concentrate than as a forage, due to its low fibre content and high protein level [17].

Agri-food by-products, whether coming from stubbles left in the field or the canning industry, constitute a supply of alternative forage for livestock, allowing the use of local resources and reducing feed costs without damaging animal performance and productivity, as long as the rations that include these feeds are balanced. The use of these by-products can also be a solution to minimise residues produced by the agro-food industry and thus reduce removal costs and emissions of polluting gases caused by uncontrolled fermentation of these agricultural wastes. In addition, the use of agro-food by-products reduces the land and supplies dedicated to the development of livestock feed, thus aiding the circular economy. However, the strong seasonality and high water content of these feeds limits their systematic use in animal feeding. Through lactic fermentation, the silage is able to conserve perishable products so that cellular respiration is suppressed, protein and vitamin degradation is prevented, and clostridial fermentation is avoided [18], reaching levels of safety that do not endanger the health of animals and do not compromise the hygienic-sanitary quality of milk or derived products.

Previous studies have shown that these by-product silages have the proper fermentative and nutritional conditions to become part of sheep and goat diets [14,19,20]. The references found in the literature about the effect of consuming these silage by-products on milk quality and composition, as well as on the health status of animals, are scarce [21–23]. None of these studies have been conducted in dairy goats, except Muelas et al. and Monllor et al. [11,24], where the effect of up to 25% inclusion of silage artichoke plant on the technological aptitude of milk was studied.

With the previous background, it is hypothesised that these by-products may be incorporated into the diet of lactating goats without detriment to their milk yield and quality. Therefore, the objective of this experiment is to study the effect of the inclusion of by-product silages (artichoke plant and broccoli by-product) in the ration of goats on milk production, macro-composition, and quality and determine the optimum level of inclusion in the ration among the three levels tested (25%, 40%, and 60%), with the aim of assuring milk nutritional quality within an integrative approach of enhanced sustainability of milk production.

2. Materials and Methods

2.1. Animals and Facilities

The animals used in this experiment were Murciano-Granadina lactating goats housed in the experimental and teaching farm of the Miguel Hernández University, Spain, with access to outdoor yards (2.30 m^2/animal), free access to water, and enough feeding space for all animals (at least 35 cm/animal and 1.50 m^2/animal as total indoor space) with a straw bed. As usual in the region, the animals were milked once a day (Casse milking parlour, $2 \times 12 \times 12$, GEA, Germany) and fed twice a day, at 8:00 a.m. and 2:00 p.m. This study was approved by the Ethical Committee of Experimentation of the Miguel Hernández University (code UMH.DTA.GRM.01.15).

2.2. Experimental Design

On the fourth month of lactation, 63 lactating goats were selected (41.2 ± 7.15 kg, 2.25 ± 0.80 kg/day, 5.39 ± 0.48 Log cell/mL). The animals were divided into seven homogeneous groups regarding body weight (BW), daily milk yield, and somatic cell count (SCC).

A short-term experiment was conducted to study the effect of inclusion in the diet of two by-product silages (artichoke plant, AP, and broccoli by-product, BB), of which their composition and fermentation quality are shown in Table 1. They were included at three levels each (25%, 40%, and 60%, expressed on a dry matter basis of the total ration); thus, seven rations were tested: 25%, 40%, and 60% of artichoke plant silage (AP25, AP40, and PAP60, respectively), the same percentages of broccoli by-product silage (BB25, BB40 and BB60), and a control diet (C), which represents the conventional ration used to feed dairy goats (alfalfa hay and a mixture of grains). Diets were formulated according to the recommendations of Fernandez et al. Ref. [25], an average amount of 2.23 kg DM/day was offered, and the seven rations were isoenergetic and isoproteic. Table 2 shows the ingredient proportion and the chemical composition of each diet. Once the pre-experimental sampling was performed, the experiment lasted 4 weeks. In the first two weeks, each group of animals adapted to their diet. In the next two weeks, data on feed consumption, milk yield, and body weight were recorded and blood and milk samples from animals were collected weekly for subsequent laboratory analyses. Bulk milk samples were collected weekly and used to determine mineral and fatty acid profile concentrations.

Table 1. Chemical composition (g/kg DM) and fermentation quality (g/kg DM) of silages included in experimental diets.

Item	BB	AP
Chemical Composition		
DM (g/kg of FM, as fed)	154	258
OM	821	828
CP	174	78.1
CF	214	296
NDF	430	571
ADF	326	374
ADL	63.4	108
EE	32.1	34.6
TP	6.73	4.96
VFA and Fermentative Metabolites		
Lactate	30.8	17.0
Acetate	117	35.2
Propionate	14.6	n.d.
Butyrate	3.80	8.56
Ethanol	14.6	3.25
Ammonia N	1.65	0.149

BB: Broccoli by-product silage; AP: Artichoke plant silage; DM: Dry matter; FM: Fresh matter; OM: Organic matter; CP: Crude protein; CF: Crude fibre; NDF: Neutral detergent fibre; ADF: Acid detergent fibre; ADL: Acid detergent lignin; EE: Ether extract; TP: Total polyphenols; VFA: Volatile fatty acids; n.d.: Not detected.

Table 2. Ingredients of experimental diets and their nutritional value.

Item	Diets						
	C	AP25	AP40	AP60	BB25	BB40	BB60
Ingredients (g/100 g DM)							
Alfalfa hay	38.0	14.7	-	-	13.5	8.50	4.60
Oat	16.0	15.0	13.0	8.0	35.0	26.5	26.6
Barley	9.50	9.00	8.00	4.51	5.50	3.72	1.23
Corn	9.08	8.43	8.00	4.35	5.16	3.60	1.19
Dried sugar beet pulp	7.36	7.00	6.50	3.53	4.18	3.00	0.960
Sunflower meal	3.36	3.12	3.00	1.61	2.00	1.33	0.440
Peas	2.50	2.32	2.09	1.20	1.42	0.990	0.330
Cottonseed	2.50	2.32	2.09	1.20	1.42	0.990	0.330
Soybean meal 44%	4.00	6.00	10.0	12.0	2.00	2.00	1.00
Corn DDGS	3.00	3.00	2.50	1.38	2.00	1.14	0.380
Sunflower seeds	2.00	1.74	2.40	1.00	1.07	0.740	0.250
Beans	1.25	1.16	1.05	0.600	1.00	0.500	0.160
Wheat	1.00	0.770	1.00	0.400	0.470	0.330	0.110
Soy hulls	0.420	0.390	0.350	0.200	0.240	0.160	0.050
Silage	-	25.0	40.0	60.0	25.0	40.0	60.0
kg DM offered/day/animal	2.24	2.26	2.20	2.30	2.22	2.21	2.20
Chemical Composition							
DM (g/kg FM)	893	554	448	361	438	334	254
g/kg DM							
OM	935	915	901	884	916	904	885
CP	162	160	163	157	162	165	169
CF	195	202	196	237	180	180	183
NDF	376	391	382	432	359	355	353
ADF	243	248	239	281	225	226	231
ADL	56.5	55.1	49.5	55.2	48.0	47.0	46.7
EE	41.9	36.5	35.1	30.5	41.3	38.5	34.7
TP	3.87	4.18	5.42	5.34	4.60	5.42	6.68
IVDMD	715	715	710	665	780	747	757
[1]ME (Mcal/kg DM)	2.37	2.30	2.29	2.19	2.39	2.36	2.29
VFA and Fermentative Metabolites (g/kg DM)							
Lactate	n.d.	14.2	23.2	24.5	33.1	41.2	56.0
Acetate	n.d.	4.91	6.04	11.9	15.1	11.0	37.8
Propionate	n.d.	n.d.	n.d.	n.d.	2.63	n.d.	4.79
Butyrate	n.d.	n.d.	n.d.	n.d.	n.d.	n.d.	n.d.
Ethanol	n.d.	1.50	1.80	1.69	9.64	12.5	23.2
Ammonia N	0.166	0.628	0.741	1.01	3.99	4.26	7.73
Fatty Acids Profile (g/100 g Total Fatty Acids)							
C6:0	0.061	0.109	0.485	0.352	0.059	0.025	0.498
C12:0	0.183	0.286	0.151	0.050	0.242	0.328	0.146
C14:0	0.440	0.502	0.413	0.357	0.542	0.539	0.465
C16:0	17.2	18.1	18.3	17.3	19.8	17.7	21.2
C16:1 c9	0.300	0.348	0.369	0.364	0.374	0.312	0.592
C18:0	3.25	3.08	2.93	3.63	2.96	3.34	2.76
C18:1 c9	26.4	25.1	22.8	31.3	30.1	34.3	21.9
C18:1 c11	1.06	1.11	1.33	1.12	2.00	2.23	3.74
C18:2n6	44.0	42.0	40.5	32.3	35.5	29.4	29.4
C18:3n3	4.07	4.79	6.75	6.43	5.79	8.18	13.0
C20:0	0.463	0.757	0.884	1.19	0.493	0.679	0.838
C20:1n9	0.323	0.408	0.300	0.336	0.464	0.386	0.423
C22:0	0.457	0.546	0.519	0.960	0.393	0.784	0.640
C24:0	0.336	0.493	0.392	0.411	0.365	0.600	0.652
SFA	23.3	24.7	26.4	26.8	25.5	24.6	29.5

Table 2. *Cont.*

Item	Diets						
	C	AP25	AP40	AP60	BB25	BB40	BB60
MUFA	28.2	27.6	26.1	33.7	33.0	37.5	27.5
PUFA	48.7	48.3	47.7	40.0	41.5	38.1	43.2
Mineral Profile							
Na (g/kg DM)	2.89	5.83	7.34	12.1	2.37	5.28	5.09
Mg (g/kg DM)	2.66	3.24	3.05	3.63	2.06	2.52	2.43
K (g/kg DM)	13.5	14.3	14.1	17.8	17.8	19.4	30.1
Ca (g/kg DM)	5.90	10.8	11.2	17.0	5.62	8.91	7.49
P (g/kg DM)	2.76	4.09	3.69	3.56	3.40	3.61	4.18
S (g/kg DM)	2.89	3.45	3.06	3.78	3.40	4.27	6.58
Se (mg/kg DM)	0.198	0.190	0.150	0.243	0.183	0.135	0.167
Zn (mg/kg DM)	49.4	44.2	41.3	34.1	43.6	42.5	36.9
Cu (mg/kg DM)	6.15	6.42	5.83	6.76	5.68	4.67	5.41
Fe (mg/kg DM)	129	414	287	460	175	161	235
Mn (mg/kg DM)	42.1	47.7	44.2	54.0	38.5	34.6	35.7

C: Control diet; AP: Diet that includes artichoke plant silage; BB: Diet that includes broccoli by-product silage; DM: Dry matter; FM: Fresh matter; OM: Organic matter; CP: Crude protein; CF: Crude fibre; NDF: Neutral detergent fibre; ADF: Acid detergent fibre; ADL: Acid detergent lignin; EE: Ether extract; TP: Total polyphenols; IVDMD: In vitro dry matter digestibility; ME: Metabolisable energy; VFA: volatile fatty acids; SFA: Saturated fatty acids; MUFA: Monounsaturated fatty acids; PUFA: Polyunsaturated fatty acids, n.d.: Not detected [1][26].

2.3. Analysed Variables

The body weight of the animals (BW, kg) was recorded by weighing them on a scale (±100 g, APC, Baxtran, Vilamalla, Spain). The feed consumption was measured twice a week and calculated by the average of the difference of the feed offered and refused on dry matter basis. The chemical composition of the silages and diets was analysed as previously described by Monllor et al. [11]. Dry matter (DM, g/kg; method 930.5), organic matter (OM, g/kg DM; method 942.05), ether extract (EE, g/kg DM; method 920.39), crude protein (CP, g/kg DM; method 984.13), and crude fibre (CF; g/kg DM; method 962.09) were determined following AOAC [27] procedures. Neutral detergent fibre (NDF, g/kg DM), acid detergent fibre (ADF, g/kg DM), and acid detergent lignin (ADL, g/kg DM) were analysed according to Van Soest et al. [28]. Total polyphenol content (TP, g/kg DM) was measured by the Folin-Ciocalteu method [29]. Volatile fatty acids (VFA, g/kg DM) (acetic, propionic, and butyric acid, also including lactic acid and ethanol) were determined by HPLC liquid chromatography (Agilent 1200, Santa Clara, CA, USA and Supelcogel C-610H column: 30 cm × 7.8 mm ID, Saint Louis, MO, USA), by Feng-Xia et al. [30] methodology. Apparent in vitro dry matter digestibility (IVDMD, g/kg DM) was measured according to Menke and Steingass [31]. Fatty acid profile analysis in diets was performed by direct methylation on the lyophilised samples, without prior extraction of the fat, according to Kramer et al. [32] and were identified by a gas chromatograph (GC-17A Shimadzu, Kyoto, Japan) coupled with a flame ionisation detector (FID) equipped with a capillary column (CP Sil 88 100 m × 0.25 mm internal diameter and 0.20 μm internal coverage, Agilent, Santa Clara, CA, USA). A mixture of fatty acid methylated esters (FAME;18912-1AMP, Sigma-Aldrich, Saint Louis, MO, USA) was used for identification of the fatty acids of the samples.

Dietary and milk minerals (Na, Mg, K, Ca, P, S, Se, Zn, Cu, Fe, and Mn) were determined by carrying out a previous digestion of the samples, according to González Arrojo et al. [6]. Microwave (MW) digestion unit Ethos Easy, Milestone (Milestone, Srl, Sorisole, Italy) equipped with a rotor for 10 TFM (chemically modified PTFE) vessels was used for sample mineralisation. The microwave program consisted of four phases (i) 5 min at 1000 W at temperatures from 100 to 60 °C; (ii) 10 min at 1000 W from 165 to 80 °C; (iii) 5 min at 1000 W from 180 to 120 °C; and, (iv) 5 min at 700 W from 180 to 120 °C. The ICP-MS (inductively coupled plasma mass spectrometry) instrument used in this study was an Agilent 7700× Octopole Reaction System (ORS) (Agilent Technologies, Tokyo, Japan). The ICP-MS operating conditions were optimised for the simultaneous determinations of all elements. ICP-MS standard solutions were prepared daily by appropriate dilution of stock standard 1000 mg/L for each element in 2% *v/v* Suprapur HNO3. An appropriate internal standard was also required for each analyte to correct physical and/or matrix interferences in ICP-MS.

The milk yield (kg/day) of every goat was determined during milking using a Lactocorder® device (Lactocorder, Balgach, Switzerland). This device collected a representative sample of 100 mL of milk at every milking of each animal for subsequent analysis. The macro-composition of milk (fat, protein, true protein, casein, whey protein, lactose, total solids, TS; non-fat total solids, NFTS; useful dry matter content, UDM, and ash; %) and urea content (mg/L) was determined by medium infrared spectroscopy equipment (MilkoScan™ FT2, Foss, Hillerød, Denmark). The SCC (10^3 cell/mL) was analysed by an electronic fluoro-optical method (DCC, DeLaval, Tumba, Sweden). Fat corrected milk yield (FCM) was calculated according to Gravert equation [33]: FCM (3.5%) = 0.433 × milk yield (kg/day) + 16.218 × fat milk yield (kg/day). Milk fatty acids were extracted by the Folch procedure, with some variations collected in Romeu-Nadal et al. [34] and were methylated following the Nudda et al. [35] method. The equipment, column, and FAME mix used for the identification of peaks of milk fatty acid profile were the same as for the diets. Atherogenicity index (AI) and thrombogenicity index (TI) were calculated according to Ulbricht and Southgate [36]. These indices provide important information because AI is related with the ability of lipids' adhesion to immunological and circulatory system cells and TI indicates the tendency to form clots in blood vessels [8]. Desaturase indices (DI) for C14:0, C16:0, and C18:0 were calculated according to Lock and Garnsworthy [37].

In order to assess the effect of the diets on goats' metabolism, blood samples were analysed. The same day as the milk sampling was performed, the fasting animals were bled and samples were collected for glucose, urea, and β-hydroxybutyrate (BHB) analysis. Blood samples were analysed with a glucose oxidase/peroxidase kit (Ref. 11503, Biosystems, Barcelona, Spain) for glucose (mg/dL), with a kinetic method (GN 10125, Gernon, Sant Joan Despí, Spain) for urea (mg/dL), and for the BHB (mmol/L), the Ranbut D-3-Hydroxybutyrate kit (RB 1007, Randox, Crumlin, UK) was used.

2.4. Calculations and Statistical Analysis

The SCC data were transformed into $\log_{10}$ scores before statistical analysis (LSCC).

BW, milk yield and macro-composition, SCC, and plasmatic profile data were performed using SAS GLIMMIX (SAS Institute Inc., Cary, NC, USA) with repeated measures, introducing the covariate of the data obtained in the pre-experimental sampling into the model and considering DIET, SAMPLING, and interaction DIET × SAMPLING as fixed effects, according to the following equation:

$$Y = \mu + Di + Sj + DixSj + covY0 + Ak + e,$$

where Y is the dependent variable, μ is the intercept, Di is the fixed effect of the diet (i = C, AP25, AP40, AP60, BB25, BB40, BB60), Sj is the fixed effect of sampling (j = 1, 2, 3), $DixSj$ is the interaction of diet with sampling, $covY0$ is the effect of the value of Y in sampling 0, Ak is the random effect of the animal, and e is the residual error. The covariance model with a lower value of the Akaike criterion (lower AIC and BIC) was used for each variable.

Milk mineral and fatty acid profile data were analysed using SAS GLM (SAS Institute Inc., Cary, NC, USA), introducing the covariate of the data obtained in the pre-experimental sampling into the model and considering DIET as a fixed effect. The level of acceptance for significance was 0.05.

3. Results

3.1. Diet Effects on Body Weight and Feed Consumption

Body weight is an indicator of the health status of the animal and optimising the inclusion of by-products involves assuring the proper health status of the goats. The treatments with the highest by-product inclusion showed a lower BW (40.2 and 38.7 kg in AP60 and BB60, respectively), while with the inclusion of 25% and 40%, no differences were observed compared to C (42.9 kg, Table 3).

Sampling and interaction Treatment × Sampling also had a significant effect on BW as an increase ($p < 0.001$) was observed in sampling 2 in treatments with 40% of by-product (+1.9 and +2.4 kg in BB40 and AP40, respectively) and then in sampling 3, they descended again. Diets were offered in a similar amount but the goats in the different treatments showed different consumptions, with group C showing the highest (2.21 kg DM/day), whereas the lowest consumption was observed in groups BB40 (1.38 kg DM/day) and BB60 (1.27 kg DM/day) compared to the other treatments, which showed intermediate consumption (AP25: 1.52, AP40: 1.54, AP60: 1.57, and BB25: 1.65 kg DM/day).

Table 3. Body weight, milk yield, and composition and SCC, according to the effects considered.

Variable	Diets								Significance		
	C	AP25	AP40	AP60	BB25	BB40	BB60	SEM	Diet	Sampling	Diet x Sampling
BW (kg)	42.9 a	41.6 ab	42.2 a	40.2 bc	41.9 ab	41.9 ab	38.7 c	0.69	***	***	***
Milk yield (kg/day)	2.24 a	2.15 ab	2.14 abc	1.92 bcd	1.90 cde	1.76 de	1.66 e	0.090	***	**	***
LSCC (Log_{10} cell/mL)	5.53	5.67	5.58	5.68	5.54	5.82	5.68	0.109	n.s	**	**
FCM (3.5%; kg/day)	2.31 ab	2.42 a	2.26 ab	2.17 abc	2.03 bc	2.00 bc	1.88 c	0.120	**	**	*
Fat (%)	3.76 b	4.25 ab	4.06 ab	4.29 ab	4.02 ab	4.25 ab	4.58 a	0.218	**	n.s.	*
Protein (%)	3.39	3.42	3.52	3.39	3.34	3.34	3.42	0.088	n.s.	n.s.	n.s.
UDM (%)	7.15 b	7.68 ab	7.59 ab	7.68 ab	7.36 ab	7.61 ab	8.01 a	0.275	*	n.s.	*
True protein (%)	3.16	3.18	3.27	3.15	3.11	3.11	3.18	0.078	n.s.	n.s.	n.s.
Casein (%)	2.68	2.69	2.76	2.66	2.65	2.65	2.72	0.061	n.s.	***	n.s.
Whey protein (%)	0.470	0.484	0.507	0.491	0.456	0.465	0.474	0.024	n.s.	***	**
Lactose (%)	4.25	4.16	4.20	4.16	4.23	4.20	4.18	0.045	n.s.	***	**
TS (%)	12.0 b	12.5 ab	12.4 ab	12.4 ab	12.2 ab	12.4 ab	12.9 a	0.28	*	*	*
NFTS (%)	8.75	8.67	8.81	8.63	8.70	8.67	8.75	0.084	n.s.	***	n.s.
Ash (%)	0.639	0.615	0.648	0.625	0.638	0.627	0.652	0.024	n.s.	n.s.	*
Milk urea (mg/L)	617 ab	587 abc	591 abc	641 a	558 bc	588 abc	542 c	23.0	**	n.s.	n.s.

C: Control diet; AP: Diet that includes artichoke plant silage; BB: Diet that includes broccoli by-product silage; 25, 40, and 60 inclusion level of by-product silage on dry matter basis %; SEM: Standard error mean; BW: Body weight; LSCC: Log_{10} somatic cell count; FCM: Fat corrected milk (3.5%); UDM: Useful dry matter content (fat + protein); TS: Total solids; NFTS: Non-fat total solids; abc: Least square means within a column having different superscripts differ significantly. * $p < 0.05$; ** $p < 0.01$; *** $p < 0.001$.

3.2. Milk Yield, Macro-Composition, and SCC

A decrease in milk yield was observed as the percentage of inclusion of by-products increased (Table 3). C, AP25, and AP40 were the treatments with the highest milk daily yield (2.24, 2.15, and 2.14 kg/day, respectively; $p < 0.001$); BB60 was associated with the lowest yield (1.66 kg/day). A tendency to decrease FCM was also observed as the percentage of inclusion of the by-product in the diet increased. The highest yield was obtained in AP25, even without significant differences compared to C or other AP treatments; BB25 and BB40 did not show significant differences compared to C, AP40, and AP60, whereas BB60 showed the lowest value. The interaction among sampling and treatments was significant as the milk yield and FCM were only significantly reduced in AP25 and AP60 during the experiment, but remained stable in the rest of the treatments.

The diet had no significant effect on LSCC. An increase of + 0.28 Log cells/mL ($p < 0.01$) was observed in AP25 between samplings 2 and 3, so that sampling and interaction with treatment were significant.

As for the macro-composition of the milk shown in Table 3, the diet only had significant effects on fat (but also affected UDM and TS) and urea (Table 3). BB60 was the one with the highest fat value and T was the lowest. The significant interaction of the treatment with the sampling in fat, UDM, TS, whey protein, and lactose was due to specific increases or decreases in sampling 2 in AP40, which returned to similar values to the previous ones at sampling 3. Both the casein content of milk and NFTS were reduced in all treatments during the experiment ($p < 0.001$). The ash content increased 0.134 percentage units in AP25 at the end of the experiment, remaining stable in the rest of the treatments. Regarding the milk urea content, AP60 was the treatment that presented the highest level (641 mg/dL; $p < 0.01$) and BB60 the lowest (542 mg/dL).

3.3. Milk Mineral Content

Milk mineral profile is shown in Table 4. Only significant differences in the Mn concentration due to dietary treatment were observed, although of small magnitude. AP40 was the treatment that presented the highest level of Mn (0.233 mg/kg DM; $p < 0.05$), followed by BB25 (0.222 mg/kg DM), whereas BB40 was the treatment showing the lowest value (0.185 mg/kg DM). These differences between treatments are not considered biologically relevant because the greatest of them, which was between AP40 and BB40, was only 0.048 mg/kg DM.

Table 4. Milk mineral profile according to the effects considered.

Mineral	Diets								
	C	AP25	AP40	AP60	BB25	BB40	BB60	SEM	Significance
Na (g/kg DM)	2.59	2.40	2.23	2.36	2.53	2.41	2.68	0.113	n.s.
Mg (g/kg DM)	0.888	0.837	0.835	0.932	0.884	0.813	0.853	0.047	n.s.
K (g/kg DM)	12.0	11.5	11.2	11.8	12.1	10.9	11.5	0.51	n.s.
Ca (g/kg DM)	8.85	7.56	8.64	8.81	8.07	7.85	7.81	0.495	n.s.
P (g/kg DM)	6.00	5.16	6.37	6.08	5.43	6.05	6.11	0.412	n.s.
S (g/kg DM)	2.45	2.29	2.44	2.45	2.35	2.40	2.37	0.107	n.s.
Se (mg/kg DM)	0.102	0.095	0.127	0.117	0.091	0.105	0.093	0.010	n.s.
Zn (mg/kg DM)	18.6	21.3	17.1	28.3	25.9	20.4	23.5	2.60	n.s.
Cu (mg/kg DM)	0.697	0.538	1.11	0.397	0.357	0.382	0.420	0.367	n.s.
Fe (mg/kg DM)	2.95	2.16	2.26	2.72	2.11	2.22	2.34	0.557	n.s.
Mn (mg/kg DM)	0.203 b	0.198 b	0.233 a	0.201 b	0.222 ab	0.185 b	0.192 b	0.010	*

C: Control diet; AP: Diet that includes artichoke plant silage; BB: Diet that includes broccoli by-product silage; 25, 40, and 60 inclusion level of by-product silage on dry matter basis %; SEM: Standard error mean. abc: Least square means within a column with different superscripts differ significantly. * $p < 0.05$.

3.4. Milk Fatty Acid Profile

Regarding the milk fatty acid profile (Table 5), some significant variations were observed, although they were quantitatively limited. Regarding the content of vaccenic acid (C18:1t11), it was observed that this was higher ($p < 0.001$) in the AP treatments, without differences compared to C. There was a higher concentration of linoleic acid (C18:2n6) in AP60 (2.53%; $p < 0.001$); however, it was at C where a higher level of other C18: 2 isomers was observed. An increase ($p < 0.001$) of α-linolenic acid (C18:3n3) was observed as the level of AP inclusion in the ration was higher and AP60 presented a higher level (0.242%). AP treatments were also those with the highest content ($p < 0.01$) in rumenic acid (CLA c9, t11), although subsequently no significant differences were found in the sum of isomers of CLA (conjugated linoleic acid) between treatments, except of BB60, of which their content was the smallest of all. Table 6 shows that as the percentage of AP inclusion increased, so did the PUFA content, and AP60 was the treatment with the highest content ($p < 0.001$) compared to all the BB treatments, without differences from C or the rest of the AP treatments. AP60 presented the highest levels ($p < 0.001$) of n3 (0.275%) and n6 (2.79%) fatty acids, the latter without differences compared to C or the other AP treatments. It also achieved the lowest ($p < 0.001$) ratio n6/n3 obtained together with BB60 (10.3 and 12.3, respectively). Regarding the lipid quality indices related to human health (AI and TI), AP40 and AP60 were the ones with the lowest value ($p < 0.001$) and therefore, were healthier. Regarding the desaturation indices of the myristic (DI14), palmitic (DI16), and stearic (DI18) fatty acids, the differences found between treatments were of small magnitude. BB60 was the one with the highest value in DI14 and DI18 (0.012% and 2.08%, respectively; $p < 0.001$) and AP60 presented a higher value of DI16 (0.061%; $p < 0.01$).

Table 5. Fatty acid composition (g/100 g total fatty acids) measured in milk according to the effects considered.

Fatty Acid	Diets							SEM	Significance
	C	AP25	AP40	AP60	BB25	BB40	BB60		
C4:0	2.21	2.66	2.53	2.57	2.53	2.62	2.67	0.586	n.s.
C6:0	3.05	3.59	3.41	3.51	3.54	3.55	3.61	0.795	n.s.
C7:0	0.052 ab	0.060 ab	0.070 ab	0.046 b	0.073 ab	0.071 ab	0.077 a	0.024	*
C8:0	4.11	4.57	4.67	4.32	4.64	4.77	4.28	0.981	n.s.
C9:0	0.065 b	0.077 ab	0.095 a	0.088 ab	0.102 a	0.102 a	0.102 a	0.023	*
C10:0	13.2	15.0	14.7	14.5	15.6	15.6	15.3	3.03	n.s.
C10:1 c9	0.037	0.040	0.033	0.036	0.047	0.036	0.034	0.017	n.s.
C11:0	0.197 ab	0.171 bc	0.186 abc	0.157 c	0.190 ab	0.201 a	0.193 ab	0.022	**
C12:0	3.23 a	2.81 bc	3.10 abc	2.66 c	3.11 abc	3.31 ab	2.93 abc	0.274	***
C12:1 c9	0.032	0.024	0.035	0.030	0.039	0.037	0.024	0.012	n.s.
iso C13:0	0.017 b	0.016 b	0.026 ab	0.028 a	0.027 a	0.016 b	0.019 ab	0.008	*
anteiso C13:0	0.025	0.025	0.030	0.030	0.030	0.031	0.026	0.008	n.s.
iso C14:0	0.055 b	0.045 b	0.060 b	0.067 ab	0.063 ab	0.058 b	0.084 a	0.019	**
C14:0	7.62 ab	7.08 ab	6.92 ab	6.74 b	7.59 ab	7.56 ab	7.76 a	0.568	*
iso C15:0	0.174 abcd	0.130 b	0.178 abc	0.184 a	0.163 abcd	0.154 bc	0.152 bcd	0.021	***
anteiso C15:0	0.226 a	0.170 c	0.208 ab	0.223 a	0.189 bc	0.181 c	0.181 c	0.021	***
C14:1 c9	0.073 bc	0.062 c	0.067 bc	0.076 abc	0.071 bc	0.080 ab	0.090 a	0.011	***
C15:0	0.652 bc	0.524 d	0.617 c	0.753 ab	0.675 bc	0.717 b	0.818 a	0.066	***
C15:1	0.070 a	0.042 d	0.048 cd	0.064 ab	0.055 bc	0.061 ab	0.055 bcd	0.011	***
iso C16:0	0.176 c	0.147 d	0.188 bc	0.225 a	0.178 c	0.204 ab	0.218 a	0.022	***
C16:0	21.5 ab	22.3 ab	20.4 ab	20.5 b	22.1 ab	22.0 ab	23.9 a	1.67	**
C16:1 t4	0.039 ab	0.003 b	0.040 ab	0.070 a	0.003 b	0.024 ab	0.048 ab	0.049	*
C16:1 t5	0.023 ab	0.005 ab	0.029 ab	0.043 a	0.000 b	0.007 ab	0.042 ab	0.036	*
C16:1 t6-7	0.105	0.089	0.112	0.139	0.097	0.060	0.085	0.148	n.s.
C16:1 t9	0.193	0.168	0.187	0.166	0.188	0.175	0.137	0.114	n.s.
C16:1 t10	0.028	0.002	0.020	0.013	0.030	0.007	0.012	0.034	n.s.
C16:1 t11-12	0.012	0.041	0.023	0.048	0.019	0.063	0.041	0.037	n.s.
C16:1 c7	0.203	0.182	0.205	0.204	0.191	0.178	0.176	0.043	n.s.
C16:1 c9	0.436 c	0.449 bc	0.491 bc	0.542 ab	0.482 bc	0.475 bc	0.617 a	0.080	**
C16:1 c10	0.029 ab	0.000 b	0.031 ab	0.047 a	0.000 b	0.012 ab	0.033 ab	0.040	*
C16:1 c11	0.000	0.002	0.004	0.006	0.000	0.003	0.011	0.009	n.s.
iso C17:0	0.249 ab	0.234 ab	0.275 a	0.223 ab	0.207 ab	0.184 b	0.165 b	0.060	**
anteiso C17:0	0.287 a	0.218 bc	0.263 ab	0.293 a	0.257 ab	0.180 c	0.282 a	0.049	***
C17:0	0.555 b	0.485 b	0.516 b	0.703 a	0.536 b	0.541 b	0.636 a	0.058	***
C17:1 c6-7	0.040	0.046	0.050	0.049	0.041	0.056	0.034	0.018	n.s.
C17:1 c8	0.000 b	0.002 b	0.000 b	0.003 b	0.002 b	0.014 b	0.035 a	0.012	***
C17:1 c9	0.104 b	0.114 b	0.121 b	0.195 a	0.119 b	0.159 a	0.215 a	0.023	***
isoC18:0	0.034 ab	0.041 ab	0.047 b	0.047 ab	0.034 b	0.057 a	0.053 ab	0.013	*
C18:0	14.1 a	12.5 ab	13.2 ab	12.2 ab	12.7 a	11.8 ab	9.9 b	0.85	***
C18:1 t4	0.068 ab	0.085 a	0.067 ab	0.049 bc	0.082 a	0.043 c	0.045 c	0.016	***
C18:1 t5	0.030 ab	0.024 b	0.031 ab	0.033 ab	0.038 a	0.017 b	0.026 ab	0.011	**
C18:1 t6-8	0.196 a	0.166 abc	0.180 ab	0.134 d	0.146 bcd	0.171 abc	0.123 cd	0.027	**
C18:1 t9	0.269 a	0.271 ab	0.245 abc	0.234 bcd	0.233 bcd	0.213 abcd	0.193 d	0.028	**
C18:1 t10	0.276 a	0.235 ab	0.230 ab	0.205 b	0.220 ab	0.235 ab	0.219 b	0.047	*
C18:1 t11	1.30 a	1.33 a	1.35 a	1.25 ab	0.98 bc	0.95 c	0.81 c	0.169	***
C18:1 t12	0.492 a	0.471 a	0.460 abc	0.396 b	0.383 bcd	0.377 bcd	0.317 d	0.049	***
C18:1 t13-14	0.059	0.000	0.058	0.000	0.062	0.114	0.037	0.117	n.s.
C18:1 c9	18.0 ab	17.6 ab	18.2 ab	19.0 a	16.3 b	16.9 ab	175 ab	1.45	*
C18:1 c11	0.043 ab	0.055 ab	0.038 ab	0.005 b	0.045 ab	0.155 a	0.052 ab	0.121	*
C18:1 c12	0.587 a	0.565 abc	0.581 a	0.536 abc	0.511 bc	0.569 ab	0.511c	0.047	**
C18:1 c13	0.124	0.116	0.112	0.115	0.115	0.119	0.112	0.019	n.s.
C18:1 c14	0.424 a	0.395 ab	0.375 ab	0.326 b	0.371 b	0.365 b	0.329 b	0.040	**
C18:1 c15	0.206	0.192	0.195	0.213	0.198	0.208	0.209	0.028	n.s.
C18:2 c9,t13	0.294 a	0.229 abc	0.246 ab	0.188 c	0.220 bc	0.220 abc	0.174 abc	0.044	**
C18:2 t8,c13	0.098 a	0.084 ab	0.083 ab	0.089 ab	0.074 b	0.089 ab	0.092 ab	0.019	*
C18:2 t9,12	0.000	0.000	0.007	0.057	0.000	0.000	0.008	0.034	n.s.
C18:2 c9,t12	0.154 a	0.117 ab	0.112 b	0.104 b	0.106 b	0.107 b	0.101 b	0.031	**
C18:2 t11,c15	0.011 ab	0.004 b	0.014 a	0.017 a	0.013 ab	0.010 b	0.017 a	0.008	**
C18:2n6	2.59 abcd	2.40 ab	2.42 ab	2.53 a	2.10 c	2.26 bc	1.98 bcd	0.193	***
C20:0	0.233 d	0.267 bc	0.280 b	0.350 a	0.237 cd	0.241 cd	0.225 d	0.029	***
C18:3n6	0.025	0.022	0.027	0.023	0.015	0.010	0.019	0.014	n.s.
C20:1 c9	0.012 ab	0.010 b	0.017 ab	0.029 a	0.000 b	0.007 b	0.008 b	0.015	**
C20:1 c11	0.038	0.050	0.053	0.049	0.052	0.053	0.040	0.018	n.s.
C18:3n3	0.183 b	0.145 c	0.152 bc	0.242 a	0.156 bc	0.179 bc	0.173 bc	0.025	***
CLA c9,t11	0.486 bc	0.510 abc	0.527 ab	0.538 ab	0.370 bc	0.377 c	0.344 bc	0.064	**
CLA t9,c11	0.044 b	0.032 c	0.038 bc	0.058 a	0.030 c	0.032 c	0.035 bc	0.009	***
CLA t10,c12	0.024	0.026	0.029	0.039	0.013	0.010	0.024	0.024	n.s.
CLA t12,14	0.017	0.012	0.023	0.025	0.009	0.006	0.022	0.017	n.s.
ΣCLA	0.528 a	0.550 a	0.549 a	0.532 a	0.529 a	0.531 a	0.482 b	0.019	***
C20:2n6	0.033	0.027	0.038	0.040	0.044	0.036	0.034	0.015	n.s.
C20:3n9	0.070 b	0.061 b	0.075 b	0.116 a	0.080 b	0.060 b	0.069 b	0.017	***
C22:0	0.023	0.027	0.019	0.025	0.018	0.021	0.027	0.015	n.s.
C20:3n3	0.000 b	0.004 b	0.013 b	0.031 a	0.000 b	0.000 b	0.000 b	0.012	***
C20:4n6	0.152 a	0.126 b	0.151 a	0.165 a	0.158 a	0.146 ab	0.153 a	0.018	***
C23:0	0.021 bc	0.019 c	0.030 abc	0.047 a	0.045 a	0.029 abc	0.038 ab	0.015	**
C20:4n3	0.001	0.001	0.001	0.001	0.010	0.001	0.001	0.009	n.s.
C22:2n6	0.000 c	0.026 b	0.001 c	0.009 bc	0.051 a	0.023 b	0.057 a	0.015	***
C24:0	0.049	0.031	0.047	0.073	0.126	0.036	0.042	0.092	n.s.

C: Control diet; AP: Diet that includes artichoke plant silage; BB: Diet that includes broccoli by-product silage; 25, 40, and 60 inclusion level of by-product silage on dry matter basis %; SEM: Standard error mean; abc: Least square means within a column having different superscripts differ significantly. * $p < 0.05$; ** $p < 0.01$; *** $p < 0.001$.

Table 6. Grouped fatty acids (g/100 g total fatty acids) and indices related to cardiovascular health and desaturation activity in milk according to the effects considered.

Variable	Diets							SEM	Significance
	C	AP25	AP40	AP60	BB25	BB40	BB60		
SFA	72.2	73.0	72.2	70.9	75.1	74.2	73.6	2.19	n.s.
MUFA	23.3	22.7	23.5	24.5	21.1	21.8	22.6	1.90	n.s.
PUFA	4.11 ab	3.86 abc	3.87 abc	4.24 a	3.40 d	3.56 cd	3.50 bcd	0.335	***
UFA	27.4	26.6	27.4	28.7	24.5	25.4	26.1	2.21	n.s.
SFA/UFA	2.64	2.77	2.64	2.50	3.10	2.95	2.85	0.326	n.s.
SCFA	22.9	26.1	25.7	24.7	26.6	26.9	25.7	5.38	n.s.
MCFA	36.2 b	35.6 b	34.3 b	34.8 b	36.5 b	36.6 b	39.4 a	2.79	*
LCFA	39.8 abc	37.4 abc	38.7 abc	41.6 ab	36.4 abc	35.4 bc	36.0 c	2.88	**
n3	0.182 b	0.151 b	0.164 b	0.275 a	0.157 b	0.178 b	0.174 b	0.034	***
n6	2.78 a	2.55 abc	2.60 ab	2.79 a	2.30 c	2.44 bc	2.18 bc	0.218	***
n6/n3	15.4 abc	17.3 ab	17.4 a	10.3 d	14.8 abc	13.8 bc	12.3 cd	2.33	***
AI	2.11 b	2.11 bc	1.95 cd	1.83 d	2.37 a	2.28 ab	2.31 ab	0.127	***
TI	3.32 b	3.30 b	3.09 cd	2.94 d	3.65 a	3.39 b	3.36 abc	0.141	***
DI C14:0	0.010 abc	0.009 abc	0.010 abc	0.011 c	0.009 abc	0.011 bc	0.012 a	0.001	***
DI C16:0	0.050 b	0.044 b	0.055 ab	0.061 a	0.044 b	0.048 b	0.050 ab	0.009	**
DI C18:0	1.55 bc	1.72 bc	1.67 b	1.80 ab	1.54 d	1.75 bc	2.08 a	0.049	***

C: Control diet; AP: Diet that includes artichoke plant silage; BB: Diet that includes broccoli by-product silage; 25, 40, and 60 inclusion level of by-product silage on dry matter basis %; SEM: Standard error mean; SFA: Saturated fatty acids; MUFA: Monounsaturated fatty acids; PUFA: Polyunsaturated fatty acids; UFA: Unsaturated fatty acids (MUFA + PUFA); SCFA: Short chain fatty acids (C6:0 a C10:0); MCFA: Medium chain fatty acids (C11:0 a C17:0); LCFA: Long chain fatty acids (C18:0 a C24:0); AI: Atherogenic index; TI: Thrombogenic index; DI: Desaturation index; abc: Least square means within a column having different superscripts differ significantly. * $p < 0.05$; ** $p < 0.01$; *** $p < 0.001$.

3.5. Plasma Metabolic Profile

Regarding the plasma metabolic profile (Table 7), it was observed that the greater the inclusion of BB in the diet, the higher the glucose level (49.5 and 50.0 mg/dL in BB40 and BB60; $p < 0.001$), although the differences were of small magnitude (42.5 mg/dL in BB25). Regarding urea, C and AP had a higher content ($p < 0.001$), while the BB treatments obtained lower levels and BB60 showed the lowest (33.2 mg/dL). The level of BHB was higher in treatments that included less by-product, such as AP25, AP40, and BB25, while it was lower in treatments that included more BB (0.299 and 0.304 mmol/L in BB40 and BB60, respectively; $p < 0.001$). There was significant interaction of treatment with sampling in the three variables due to the different behaviour throughout the experiment between treatments: Glucose increased ($p < 0.001$) with the progress of the experiment in all treatments except BB60; blood urea was reduced ($p < 0.001$) at sampling 2 in BB25 and BB40 and increased at sampling 3 in BB25, BB40, and BB60; BHB increased ($p < 0.01$) at the end of the experiment in BB25, BB60, and AP60, while in C, BB40, AP25, and AP40 remained stable.

Table 7. Plasmatic profile according to the effects considered.

Variable	Diets							SEM	Significance		
	C	AP25	AP40	AP60	BB25	BB40	BB60		Diet	Sampling	Diet x Sampling
Glucose (mg/dL)	44.6 bc	47.7 ab	45.0 bc	48.3 ab	42.5 c	49.5 a	50.0 a	1.52	***	***	***
Plasma urea (mg/dL)	52.0 a	50.7 a	50.9 a	49.2 a	38.8 bc	39.8 b	33.2 c	2.14	***	**	***
BHB (mmol/L)	0.336 bc	0.522 a	0.424 ab	0.376 bc	0.421 ab	0.299 c	0.304 c	0.040	***	n.s.	**

C: Control diet; AP: Diet that includes artichoke plant silage; BB: Diet that includes broccoli by-product silage; 25, 40, and 60 inclusion level of by-product silage on dry matter basis %; SEM: Standard error mean; BHB: β-hydroxybutyrate; abc: Least square means within a column having different superscripts differ significantly. ** $p < 0.01$; *** $p < 0.001$.

4. Discussion

4.1. Diet Effects on Body Weight and Feed Consumption

One of the factors that affects the total volume of the diet and its consumption by livestock is the moisture content, as Jackson and Forbes [38] pointed out. This effect is especially important in the short term as herbivores are able to progressively modify the volume of the rumen to increase the speed of transit of the digesta [39], so in the long term, this effect would have less influence. In this experiment, carried out in the short term, diet C was the one presenting the highest DM content and feed consumption (2.21 kg DM/day). On the contrary, diets BB40 and BB60 contained a greater amount of water and were bulkier and presented less consumption. In addition, diets with silage showed higher concentrations of VFA and other substances resulting from fermentation. The presence of propionic acid in BB60 (4.79 g/kg DM), as well as a higher concentration of ammonia N in both BB40 and BB60, also occurred in treatments with lower consumption due to the depressing effect on feed consumption demonstrated by Baumont [40]. The feed consumption of the BB treatments was superior to those found by Meneses [41] (0.508 kg DM/day) in Murciano-Granadina castrated males, whose ration incorporated 55% of BB silage. All BW values were normal for the Murciano-Granadina breed [42,43]. The greatest reduction in BW was in BB60, as well as the greatest reduction in feed consumption (1.27 kg DM/day and 38.7 kg).

4.2. Milk Yield, Macro-Composition, and SCC

The treatments that presented a higher feed consumption were those that had a higher milk yield. The values obtained are similar to the yield obtained with the equation proposed by León et al. Ref. [44] for the modelling of the Murciano-Granadina lactation curve, which stands at 1.93 kg/day between the fourth and fifth lactation months, which is where the animals used in this experiment were located. The highest percentage of fat in BB60 (4.59%) was probably due to a concentration effect (being the treatment with the lowest yield) and its highest content in acetic acid (37.8 g/kg DM, triple the rest) in the diet, which is an extra-lipogenic nutrient precursor of fat synthesis. Van Knegsel et al. [45] observed similar effects in dairy cows when part of the corn in the diet was replaced by beet pulp. Due to a higher fat concentration in BB60, UDM and TS also reached the highest values in this treatment (8.03% and 12.9%, respectively), exceeding C by almost a percentage point. The urea level of all treatments was found to be within the optimal range for goats recommended by the Interprofessional Dairy Laboratory of Castilla-La Mancha (LILCAM), which is between 500 and 700 mg/L. The differences found in the milk urea content can be explained by the different levels of feed consumption of the treatments. BB60 presented less feed consumption, in particular refusing part of the offered BB, which probably induced lower total protein intake and lower levels of milk urea, as Jimeno et al. [46] noticed.

4.3. Milk Mineral Content

The macromineral values correspond to those found by Mellado and García [47] in goat crossings. The composition of the diet of animals largely determines the concentrations of minerals in milk [48]. As there were no large differences in the content of the different minerals in the diets, no significant differences were subsequently observed in the milk of the different treatments, which is important for the technological aptitude of the milk, given the relevance of Ca and P in the setting and development of the microstructure of cheese [49], the main destination of goat's milk. Only the Mn had a higher concentration in AP40 (0.233 mg/kg DM), although with such tight differences that they are not biologically relevant.

4.4. Milk Fatty Acid Profile

The milk of animals fed with AP60 had a higher content of n3 fatty acids, which caused a lower n6/n3 ratio, which is positive for the prevention of coronary and cardiovascular diseases [50]. On the other hand, C, AP25, AP40, and AP60, of which their diets had the highest levels of PUFA, were the treatments with milk richest in vaccenic, rumenic, and PUFA, as reported by Collomb et al. [51], who observed differences in the PUFA and vaccenic content in the milk of cows fed with high mountain pastures and in lowland plains because the plants that made up the mountain meadows had a higher concentration of PUFA.

Another factor that could influence the increase of PUFA in AP treatments was the slightly higher content of total polyphenols (TP) in the diet, although lower than that of BB60. However, the lower feed consumption of BB60 could mean that the total TP intake does not reach those of the AP treatments. Several studies have demonstrated the inhibitory action of dietary polyphenols on ruminal biohydrogenation of PUFA, without detrimental effects on milk yield and composition, due to interference with microbial flora [52–55]. These effects have also been observed in sheep with small amounts in the diet of by-products rich in TP [56,57]. Cappucci et al. Ref. [9] also observed how after increasing the TP content of the diet of Comisana sheep by including different levels of olive by-product, the concentration of linoleic (C18:2n6) and α-linolenic (C18:3n3) in milk was increased.

As a result of a lower milk content of C12:0, C14:0, C16:0, and C18:0, AP40 and AP60 had the lowest levels of AI and TI, so the milk of these animals would be of higher quality in terms of human health [42]. The values obtained from AI in all the treatments of this study are below those found by Molina-Alcaide et al. Ref. [42] in Murciano goats fed with conventional ration supplemented with feed blocks of olive by-products. The desaturation indices obtained in this experiment are similar to those provided by Baldin et al. Ref. [58] in a study in goats that received a dietary CLA supplement.

4.5. Plasma Metabolic Profile

Despite the differences found in the metabolic profile of the different treatments, glucose, urea, and BHB levels remained within the ranges considered optimal for goats [59], except for the urea value in BB60, which was slightly lower. As Friggens et al. [60] observed in goats' performance, the level of BHB was generally low and particularly in those treatments showing lower feed consumption (BB40 and BB60) because goats, as lactating animals, adapt their milk yield to the level of feed intake, as seen in Table 3. This reduces the metabolic load and allows them to maintain adequate body reserves turnover. Due to the strong relationship between plasma and milk urea content [61], the lower levels of blood urea were found in the same treatments with the lowest values of milk urea.

5. Conclusions

The findings of this study highlighted that a threshold level of AP or BB inclusion in dairy goat diets, without negative effects on milk yield, composition, mineral and fatty acid profile, as well as metabolic status of the animals, would be 40% of the dietary dry matter.

The inclusion of artichoke plant and broccoli by-product silages in high doses (60%) caused lower feed consumption and lower milk yield. Inclusion at 60% of AP and BB increased the milk TS, although not enough to compensate for the reduced yield, resulting in lower FCM in the case of BB60. No differences were found regarding the milk mineral profile. Inclusion of the artichoke plant silage in the animals' diet improved the milk lipid profile from the point of view of human health (AI, TI) compared to broccoli silage, due to a lower SFA content (C12:0, C14:0, and C16:0) and a higher concentration of PUFA, especially vaccenic acid (C18:1 trans11) and rumenic acid (CLA cis9, trans11), without any differences compared to the control treatment. Regarding sanitary status, the plasma metabolic profile in broccoli treatments reflects that goats ate grains and alfalfa, whereas broccoli was the last ingredient, impairing its consumption, especially at the high concentration (60%). In addition, the diets that included 60% of by-product silages caused a reduction in BW.

Author Contributions: Conceptualisation, G.R. and J.R.D.; Data curation, G.R., A.J.A.-B., and J.R.D.; Methodology, P.M., G.R., A.R., E.S., and J.R.D.; Resources, G.R.; Writing—original draft, P.M.; Writing—review & editing, G.R., A.S.A., C.A.S.-C., E.S., and J.R.D. All authors have read and agreed to the published version of the manuscript.

Funding: This research received funding from the Spanish Ministry of Economy, Industry and Competitiveness and the European Regional Development Fund, grant number AGL2015-64518-R (MINECO/FEDER, UE). Paula Monllor was funded by an FPU grant (Reference number: FPU14/06058) from the Spanish Ministry of Education.

Acknowledgments: Aprovertia S.L., a technology-based company of Miguel Hernández University of Elche, provided the facilities for manufacturing the silages.

Conflicts of Interest: The authors declare no conflict of interest. The funders had no role in the design of the study; in the collection, analyses, or interpretation of the data; in the writing of the manuscript, or in the decision to publish the results.

References

1. Guo, M. Goat milk. In *Encyclopedia of Food Sciences and Nutrition*, 2nd ed.; Caballero, B., Finglas, P., Toldra, F., Eds.; Academic Press: Cambridge, MA, USA, 2003; pp. 2944–2949.

2. Pulina, G.; Milán, M.; Lavín, M.; Theodoridis, A.; Morin, E.; Capote, J.; Thomas, D.; Francesconi, A.; Caja, G. Invited review: Current production trends, farm structures, and economics of the dairy sheep and goat sectors. *J. Dairy Sci.* **2018**, *101*, 6715–6729. [CrossRef]

3. Turck, D. Cow's milk and goat's milk. *World Rev. Nutr. Diet* **2013**, *108*, 56–62. [PubMed]

4. Chen, L.; Li, X.; Li, Z.; Deng, L. Analysis of 17 elements in cow, goat, buffalo, yak, and camel milk by inductively coupled plasma mass spectrometry (ICP-MS). *RSC Adv.* **2020**, *10*, 6736–6742. [CrossRef]

5. Djordjevic, J.; Ledina, T.; Baltic, M.Z.; Trbovic, D.; Babic, M.; Bulajic, S. Fatty acid profile of milk. *IOP Conf. Ser. Earth Environ. Sci.* **2019**, *333*, 012057. [CrossRef]

6. González-Arrojo, A.; Soldado, A.; Vicente, F.; Fernández Sánchez, M.L.; Sanz-Medel, A.; de la Roza-Delgado, B. Changes on levels of essential trace elements in selenium naturally enriched milk. *J. Food Nutr. Res.* **2016**, *4*, 303–308.

7. Halmemies-Beauchet-Filleau, A.; Shingfield, K.; Simpura, I.; Kokkonen, T.J.; Jaakkola, S.; Toivonen, V.; Vanhatalo, A. Effect of incremental amounts of camelina oil on milk fatty acid composition in lactating cows fed diets based on a mixture of grass and red clover silage and concentrates containing camelina expeller. *J. Dairy Sci.* **2017**, *100*, 305–324. [CrossRef]

8. Hilali, M.; Rischkowsky, B.; Iniguez, L.; Mayer, H.K.; Schreiner, M. Changes in the milk fatty acid profile of Awassi sheep in response to supplementation with agro-industrial by-products. *Small Rumin. Res.* **2018**, *166*, 93–100. [CrossRef]

9. Cappucci, A.; Alves, S.P.; Bessa, R.J.; Buccioni, A.; Mannelli, F.; Pauselli, M.; Viti, C.; Pastorelli, R.; Roscini, V.; Serra, A.; et al. Effect of increasing amounts of olive crude phenolic concentrate in the diet of dairy ewes on rumen liquor and milk fatty acid composition. *J. Dairy Sci.* **2018**, *101*, 4992–5005. [CrossRef]

10. Schulz, F.; Westreicher-Kristen, E.; Molkentin, J.; Knappstein, K.; Susenbeth, A. Effect of replacing maize silage with red clover silage in the diet on milk fatty acid composition in cows. *J. Dairy Sci.* **2018**, *101*, 7156–7167. [CrossRef]

11. Monllor, P.; Romero, G.; Sendra, E.; Atzori, A.S.; Diaz, J.R. Short-term effect of the inclusion of silage Artichoke by-products in diets of dairy goats on milk quality. *Animals* **2020**, *10*, 339. [CrossRef]

12. Caputo, A.R.; Morone, G.; Di Napoli, M.A.; Rufrano, D.; Sabia, E.; Paladino, F.; Sepe, L.; Claps, S. Effect of destoned olive cake on the aromatic profile of cows' milk and dairy products: Comparison of two techniques for the headspace aroma profile analysis. *Ital. J. Agron.* **2015**, *10*, 15. [CrossRef]

13. FAO. Food and Agriculture Organization of the United Nations. Available online: http://www.fao.org/faostat/en/#data/FO (accessed on 4 April 2019).

14. Hernández, F.; Pulgar, M.A.; Cid, J.M.; Moreno, R.; Ocio, E. Nutritive assessment of artichoke crop residues (Cynara scolymus L): Sun dried leaves and whole plant silage. *Arch. Zootec* **1992**, *41*, 257–264.

15. Wernli, C.; Thames, I. Utilization of fodder residue of artichoke (Cynara scolymus L.) as silage. I. Factors affecting its conservation. *Av. Prod. Anim.* **1989**, *14*, 79–89.

16. Ros, M.; Pascual, J.A.; Ayuso, M.; Morales, A.B.; Miralles, J.R.; Solera, C. *Estrategias Sostenibles para un Manejo Integral de los Residuos y Subproductos Orgánicos de la Industria Agroalimentaria. Proyecto Life+ Agrowaste*; CEBAS-CSIC, CTC y AGRUPAL: Murcia, España, 2012.

17. Wiedenhoeft, M.H.; Barton, B.A. Management and environment effects on brassica forage quality. *Agron. J.* **1907**, *86*, 227–232. [CrossRef]

18. Shinners, K.J.; Wepner, A.D.; Muck, R.E.; Weimer, P.J. Aerobic and anaerobic storage of single-pass, chopped corn stover. *BioEnergy Res.* **2010**, *4*, 61–75. [CrossRef]

19. Meneses, M.; Megías, M.D.; Madrid, J.; Martínez-Teruel, A.; Hernández, F.; Oliva, J. Evaluation of the phytosanitary, fermentative and nutritive characteristics of the silage made from crude artichoke (Cynara scolymus L.) by-product feeding for ruminants. *Small Rumin. Res.* **2007**, *70*, 292–296. [CrossRef]

20. Monllor, P.; Muelas, R.; Roca, A.; Sendra, E.; Romero, G.; Díaz, J.R. Nutritive and fermentative evaluation of silages made from plant of artichoke and artichoke and broccoli by-product. In *Proceedings of the XLII Nationas and XVIII International Congress of Spanish Society of Sheep and Goat Husbandry (SEOC), Salamanca, Spain, 20–22 September 2017*; Spanish Society of Sheep and Goat Husbandry: Sevilla, Spain, 2017; pp. 139–145.

21. Marsico, G.; Ragni, M.; Vicenti, A.; Jambrenghi, A.C.; Tateo, A.; Giannico, F.; Vonghia, G. The quality of meat from lambs and kids reared on feeds based on Artichoke (Cynara Scolymus L.) bracts. *Acta Hortic.* **2005**, *681*, 489–494. [CrossRef]

22. Jaramillo, D.; Buffa, M.; Rodríguez, M.; Pérez-Baena, I.; Guamis, B.; Trujillo, A.-J. Effect of the inclusion of artichoke silage in the ration of lactating ewes on the properties of milk and cheese characteristics during ripening. *J. Dairy Sci.* **2010**, *93*, 1412–1419. [CrossRef]

23. Salman, F.M.; El-Nomeary, Y.A.A.; Abedo, A.A.; Abd El-Rahman, H.H.; Mohamed, M.I.; Ahmed, S.M. Utilization of artichoke (Cynara scolymus) by-products in sheep feeding. *Am.-Eurasian J. Agric. Environ. Sci.* **2014**, *14*, 624–630.

24. Muelas, R.; Monllor, P.; Romero, G.; Sayas-Barberá, E.; Navarro, C.; Diaz, J.R.; Sendra, E. Milk technological properties as affected by including Artichoke by-products silages in the diet of dairy goats. *Foods* **2017**, *6*, 112. [CrossRef]

25. Fernández, C.; Sánchez-Séiquer, P.; Navarro, M.J.; Garcés, C. Modeling the voluntary dry matter intake in Murciano-Granadina dairy goats. In *Sustainable Grazing, Nutritional Utilization and Quality of Sheep and Goat Products*; Molina, A.E., Ben, S.H., Biala, K., Morand-Fehr, P., Eds.; CIHEAM: Zaragoza, Spain, 2005; pp. 395–399.

26. INRA. *Alimentation des Bovins, Ovins et Caprins*; Jarrige, R., Ed.; INRA: Paris, France, 1988; p. 471.

27. AOAC. *Official Methods of Analysis*, 16th ed.; Cunniff, P., Ed.; Association of Official Analytical Chemists: Washington, WA, USA, 1999.

28. Van Soest, P.J.; Robertson, J.B.; Lewis, B.A. Methods for dietary neutral detergent fibre and nonstarch polysacacharides in relation to animal nutrition. *J. Dairy Sci.* **1991**, *74*, 3583–3597. [CrossRef]

29. Kim, D.; Jeong, S.W.; Lee, C.Y. Antioxidant capacity of phenolic phytochemicals from various cultivars of plums. *Food Chem.* **2003**, *81*, 321–326. [CrossRef]

30. Liu, F.-X.; Fu, S.-F.; Bi, X.-F.; Chen, F.; Liao, X.-J.; Hu, X.-S.; Wu, J. Physico-chemical and antioxidant properties of four mango (Mangifera indica L.) cultivars in China. *Food Chem.* **2013**, *138*, 396–405. [CrossRef] [PubMed]

31. Menke, K.H.; Steingass, H. Estimation of the energetic feed value obtained from chemical analysis and in vitro gas production using rumen fluid. *Anim. Res.* **1988**, *23*, 103–116.

32. Kramer, J.K.G.; Fellner, V.; Dugan, M.E.R.; Sauer, F.D.; Mossoba, M.M.; Yurawecz, M.P. Evaluating acid and base catalysts in the methylation of milk and rumen fatty acids with special emphasis on conjugated dienes and total trans fatty acids. *Lipids* **1997**, *32*, 1219–1228. [CrossRef]

33. Gravert, H.O. *Dairy Cattle Production*; Elsevier Science: New York, NY, USA, 1987; p. 234.

34. Romeu-Nadal, M.; Morera-Pons, S.; Casteltratamiento, A.I.; López-Sabater, M.C. Comparison of two methods for the extraction of fat from human milk. *Anal. Chim. Acta* **2004**, *513*, 457–461. [CrossRef]

35. Nudda, A.; McGuire, M.; Battacone, G.; Pulina, G. Seasonal variation in conjugated linoleic acid and vaccenic acid in milk fat of sheep and its transfer to cheese and ricotta. *J. Dairy Sci.* **2005**, *88*, 1311–1319. [CrossRef]

36. Ulbricht, T.; Southgate, D. Coronary heart disease: Seven dietary factors. *Lancet* **1991**, *338*, 985–992. [CrossRef]

37. Lock, A.; Garnsworthy, P. Seasonal variation in milk conjugated linoleic acid and Δ9-desaturase activity in dairy cows. *Livest. Prod. Sci.* **2003**, *79*, 47–59. [CrossRef]

38. Jackson, N.; Forbes, T.J. The voluntary intake by cattle of four silages differing in dry matter content. *Anim. Sci.* **1970**, *12*, 591–599. [CrossRef]

39. Schettini, M.A.; Prigge, E.C.; Nestor, E.L. Influence of mass and volume of ruminal contents on voluntary intake and digesta passage of a forage diet in steers. *J. Anim. Sci.* **1999**, *77*, 1896–1904. [CrossRef] [PubMed]

40. Baumont, R. Palatabilité et comportement alimentaire chez le ruminant. *INRA Prod. Anim.* **1996**, *9*, 349–358.

41. Meneses, M. Evaluación Nutritiva y Fermentativa del Ensilado de dos Subproductos Agroindustriales, Brócoli (Brassica oleracea, L. var. Itálica) y Alcachofa (Cynara Scolymus, L) para su Empleo en la Alimentación Animal. Ph.D. Thesis, University of Murcia, Murcia, Spain, 2002.

42. Molina-Alcaide, E.; Morales-García, E.; Martín-García, A.; Ben Salem, H.; Nefzaoui, A.; Sanz-Sampelayo, M.; Morales-García, Y.E. Effects of partial replacement of concentrate with feed blocks on nutrient utilization, microbial N flow, and milk yield and composition in goats. *J. Dairy Sci.* **2010**, *93*, 2076–2087. [CrossRef] [PubMed]

43. Fernández, C.; Pérez-Baena, I.; Marti, J.; Palomares, J.; Jorro-Ripoll, J.; Segarra, J. Use of orange leaves as a replacement for alfalfa in energy and nitrogen partitioning, methane emissions and milk performance of murciano-granadina goats. *Anim. Feed. Sci. Technol.* **2019**, *247*, 103–111. [CrossRef]

44. León, J.; Macciotta, N.P.; Gama, L.T.; Barba, C.; Delgado, J. Characterization of the lactation curve in Murciano-Granadina dairy goats. *Small Rumin. Res.* **2012**, *107*, 76–84. [CrossRef]

45. Van Knegsel, A.T.; Brand, H.V.D.; Dijkstra, J.; Van Straalen, W.; Heetkamp, M.; Tamminga, S.; Kemp, B. Dietary energy source in dairy cows in early lactation: Energy partitioning and milk composition. *J. Dairy Sci.* **2007**, *90*, 1467–1476. [CrossRef]

46. Jimeno, V.; Rebollar, P.G.; Castro, T. Nutrición y alimentación del caprino de leche en sistemas intensivos de explotación. In Proceedings of the Alimentación Práctica del Caprino de Leche en Sistemas Intensivos. XIX Curso de Especialización FEDNA, Madrid, Spain, 23–24 October 2003; pp. 155–178.

47. Mellado, M.; García, J. Effects of abortion and stage of lactation on chemical composition and mineral content of goat milk from mixed-breed goat on rangeland. *APCBEE Procedia* **2014**, *8*, 1–5. [CrossRef]

48. Rey-Crespo, F.; Miranda, M.; López-Alonso, M. Essential trace and toxic element concentrations in organic and conventional milk in NW Spain. *Food Chem. Toxicol.* **2013**, *55*, 513–518. [CrossRef]

49. Pastorino, A.; Hansen, C.; McMahon, D. Effect of pH on the chemical composition and structure-function relationships of cheddar cheese. *J. Dairy Sci.* **2003**, *86*, 2751–2760. [CrossRef]

50. Hu, F.B. Optimal diets for prevention of coronary heart disease. *JAMA* **2002**, *288*, 2569–2578. [CrossRef]

51. Collomb, M.; Bisig, W.; Bütikofer, U.; Sieber, R.; Bregy, M.; Etter, L. Fatty acid composition of mountain milk from Switzerland: Comparison of organic and integrated farming systems. *Int. Dairy J.* **2008**, *18*, 976–982. [CrossRef]

52. Castro-Carrera, T.; Toral, P.G.; Frutos, P.; McEwan, N.R.; Hervás, G.; Abecia, L.; Pinloche, E.; Girdwood, S.; Belenguer, A. Rumen bacterial community evaluated by 454 pyrosequencing and terminal restriction fragment length polymorphism analyses in dairy sheep fed marine algae. *J. Dairy Sci.* **2014**, *97*, 1661–1669. [CrossRef] [PubMed]

53. Buccioni, A.; Pauselli, M.; Viti, C.; Minieri, S.; Pallara, G.; Roscini, V.; Rapaccini, S.; Trabalza-Marinucci, M.; Lupi, P.; Conte, G.; et al. Milk fatty acid composition, rumen microbial population, and animal performances in response to diets rich in linoleic acid supplemented with chestnut or quebracho tannins in dairy ewes. *J. Dairy Sci.* **2015**, *98*, 1145–1156. [CrossRef] [PubMed]

54. Costa, M.; Alves, S.P.; Cabo, Â.; Guerreiro, O.; Stilwell, G.; Dentinho, M.T.; Bessa, R.J.B. Modulation ofin vitrorumen biohydrogenation byCistus ladanifertannins compared with other tannin sources. *J. Sci. Food Agric.* **2016**, *97*, 629–635. [CrossRef] [PubMed]

55. Correddu, F.; Fancello, F.; Chessa, L.; Atzori, A.; Pulina, G.; Nudda, A. Effects of supplementation with exhausted myrtle berries on rumen function of dairy sheep. *Small Rumin. Res.* **2019**, *170*, 51–61. [CrossRef]

56. Nudda, A.; Correddu, F.; Atzori, A.; Marzano, A.; Battacone, G.; Nicolussi, P.; Bonelli, P.; Pulina, G. Whole exhausted berries of Myrtus communis L. supplied to dairy ewes: Effects on milk production traits and blood metabolites. *Small Rumin. Res.* **2017**, *155*, 33–38. [CrossRef]

57. Nudda, A.; Buffa, G.; Atzori, A.S.; Cappai, M.G.; Caboni, P.; Fais, G.; Pulina, G. Small amounts of agro-industrial byproducts in dairy ewes diets affects milk production traits and hematological parameters. *Anim. Feed. Sci. Technol.* **2019**, *251*, 76–85. [CrossRef]

58. Baldin, M.; Dresch, R.; De Souza, J.; Fernandes, D.; Gama, M.; Harvatine, K.; Oliveira, D. CLA induced milk fat depression reduced dry matter intake and improved energy balance in dairy goats. *Small Rumin. Res.* **2014**, *116*, 44–50. [CrossRef]

59. Rivas, J.; Rossini, M.; Colmenares, O.; Salvador, A.; Morantes, M.; Valerio, D. Effect of feeding on the profile metabolic goats in canary in the tropics. In Proceedings of the 4th Simposium of the Latinomerican Asociation in Animal Science, Quevedo, Ecuador, 13–15 November 2014; pp. 125–132.

60. Friggens, N.; Duvaux-Ponter, C.; Etienne, M.; Mary-Huard, T.; Schmidely, P. Characterizing individual differences in animal responses to a nutritional challenge: Toward improved robustness measures. *J. Dairy Sci.* **2016**, *99*, 2704–2718. [CrossRef]

61. Bonanno, A.; Di Grigoli, A.; Di Trana, A.; Di Gregorio, P.; Tornambè, G.; Bellina, V.; Claps, S.; Maggio, G.; Todaro, M. Influence of fresh forage-based diets and αS1-casein (CSN1S1) genotype on nutrient intake and productive, metabolic, and hormonal responses in milking goats. *J. Dairy Sci.* **2013**, *96*, 2107–2117. [CrossRef]

MDPI
St. Alban-Anlage 66
4052 Basel
Switzerland
Tel. +41 61 683 77 34
Fax +41 61 302 89 18
www.mdpi.com

Foods Editorial Office
E-mail: foods@mdpi.com
www.mdpi.com/journal/foods

9 783036 543178